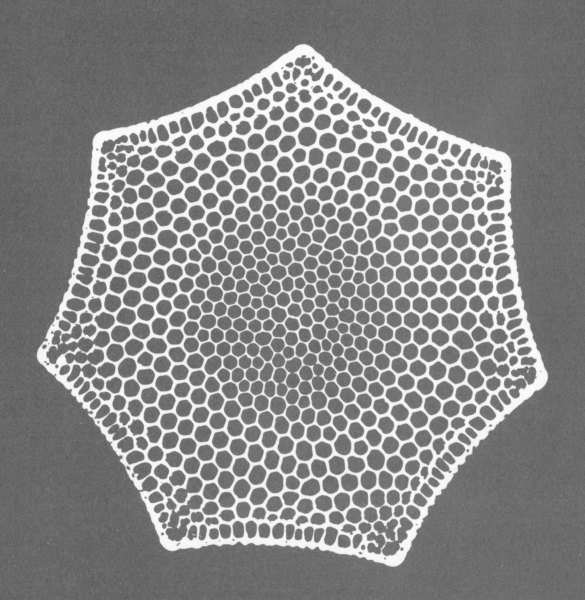
KU-726-095

the
life-giving
sea

the
life~giving
sea

David Bellamy

HAMISH HAMILTON
LONDON

First published in Great Britain 1975
by Hamish Hamilton Limited
90 Great Russell Street London WC1

© David Bellamy 1975

All rights reserved. No part of this publication may be reproduced,
stored in a retrieval system or transmitted in any form or by any
means, electronic, mechanical, photocopying, recording or
otherwise, without the prior permission of the publisher

Produced by Walter Parrish International Limited, London
Designed by Judy A Tuke

Printed and bound in Spain by Roner SA, Madrid
Dep. Legal: SS 337–1975

ISBN 0 241 89233 3

to Fluff

Preface

In 1859 Charles Darwin published his theory of evolution under the title *On the Origin of Species by means of Natural Selection*. What made it a best seller was the sub-title 'The preservation of favoured races in the struggle for life'. This greatly appealed to the Victorians because they understood the meaning of the word struggle and certainly considered themselves the favoured race, and thus arose the catch phrase 'survival of the fittest in the struggle for existence'.

It could not have come at a more opportune time. The developing industrial society was turning away from religion and was looking for a new philosophy on which to pin its hopes or, at the very least, blame its failures. The new dogma of evolution not only filled this gap but became, unwittingly perhaps, the cornerstone of the rat race philosophy of the evolving society. To be one of the 'fittest' was the new ideal, and the meaning of survival became lost in the euphoria of success. But recent developments relating to man's activities on a world scale have cast doubts on the benefits of this success, on the fitness of man in his environment, and hence on the prospects of his survival.

In this book I hope to present sufficient information to allow the reader to 'think' ecology; to understand why there are, and must be, such an enormous variety of plant and animal species; to appreciate the importance of conservation; and hence to discover for himself the true philosophy of evolutionary fitness. In essence this discards the rat race philosophy of the super organism and replaces it with that of the evolution of integrated systems of organisms which, working together, not only survive but reap the fullest potential of the environments of this earth; the evolution not of organisms, but of living systems in which interdependence is the key. And nowhere is this more apparent than in the sea— that vast resource covering 71 per cent of this ill-named planet, Earth, about which we know very little.

Man's insatiable quest for adventure and knowledge has already taken him into outer space to probe the resources of other planets. But it is fair to say that the likelihood of our being able to use these resources, thus cashing in on the enormous investment in space research, is decades away. Futhermore, the present economic climate may well curtail the space programme altogether. Yet the continental shelves, which are at a maximum only 200 metres away from our atmosphere, represent a land mass the size of Africa, whose potential is waiting to be discovered and used.

The world's deposit account of energy and raw materials (namely the fossil fuels) is being rapidly used up, and there are now doubts concerning the safety of atomic power. Even if the deposits yet to be discovered under the sea come up to our wildest expectations, the time is not too far away when man's endeavours will be more and more dependent on the current account of daily solar input both for safe energy and organic raw materials. Tidal power, wave power, current power, and the thermal difference between the warm surface waters and the cold of the abyss— these all represent energy for the taking on a massive scale.

The oceans are not only the world's final resource; they are also the world's final sink, a sink for all the products of global erosion, both natural and man-made. The problems pertaining to marine pollution and marine resource interaction must be researched, debated, and the necessary legislation agreed upon. What is perhaps most exciting is that the oceans belong to no nation and yet connect all of them. They are an international resource whose development must be a cooperative effort; an effort that could draw the nations closer together; an effort that must not be allowed to push them further apart.

In 1873 Anton Dohrn founded the Stazione Zoologica in Naples. His aim was to put the Darwinian theory to the test in the marine environment. He created a unique place of work, with the living laboratory of the Mediterranean at its doorstep and an extension of it within the great aquariums of the institute. Here many of the best brains of the world found the potential necessary to explore together the wonders of evolution in the sea.

It is the knowledge of the unique sense of cooperative research (working together to a common goal) which evolved in the Naples marine biological station that makes me believe man can solve his many problems and will gain the stamp of evolutionary success.

David Bellamy

March 7th 1874

My dear Dr Dohrn

I have just heard from Huxley
that you are much over-worked.
and troubled about the Zoological
Stations This This has grieved me
much. I am glad that you are
now willing to receive assistance
from English naturalists, not
on your own account, but for
the Zoological Station. I have

written to Hu[xley]
communicat[ing]
who will be
aid . in this
do not know
may not be
present, I ha[ve]
would allow
once. my sub[scription]
and one of [my]
two sons [George?]
As I want

about ~~sende~~
~~…~~tte any mew
~~…~~y to give their
work. As I
~~w~~hether you may
~~…~~t of money at
thought that you
~~…~~t send you at
~~…~~plica of £ 100,
each from my
~~…~~ & Francis. ✗
~~…~~atch to days

post, I write in haste, but believe
me you have my heartfelt symp
= pathy & respect.
        Yours very sincerely
        Charles Darwin

Dr Anton Dohrn

Letter from Charles Darwin to Anton Dohrn, founder
of the Stazione Zoologica (marine biological station) in
Naples.

Contents

potential to man?

Author's acknowledgements

I would like to record my thanks to all the plants and animals I have encountered while on my short excursions under the sea, and to all the people who have made these dives possible and have given me the knowledge to understand all that I have seen. I can only mention a few, and each will, I hope, forgive me for linking their names with this inadequate book: Thomas J. Bellamy, E. F. Hutchings, John Clegg, George Fluck, Ned Norris, Comyns J. A. Berkeley, C. C. Henchel, Eleanor Brown, Muriel Sutton, Ed Cossins, Stanislaw Kulczynski, John Waughman, Jim Barnes, Tom Wright, Dai Jones, Ed Drew, John Lythgoe, Lee Kenyon, Dickie Bird, Alan Baldwin, Peter Edwards, Tony Johnson, Mont Hirons, and Peter Dohrn.

water, energy
and life

Water and the origin of the oceans

Of all the matter that exists on the surface of our planet, water is the commonest. It covers 70 per cent of the earth to an average depth of 4000 metres. It is the substance of streams, rivers, ponds, oceans, clouds, and rain. It is found combined in rocks and minerals that are as old as the earth itself, and it constitutes more than half of all living matter (which has only appeared on earth in comparatively recent times). Not only is it the commonest of all substances but it is one of the most peculiar, about which we know very little.

In the simplest chemical terms, water is a compound of hydrogen and oxygen: hydrogen is a gas that burns very easily and oxygen is a gas that supports combustion. The best way to make water is to burn hydrogen gas in an atmosphere of pure oxygen. The combination can be explosively spectacular; the result—just plain water. However, this unique combination has some remarkable properties based on the fact that water molecules attract one another and thus form themselves into cohesive masses.

Above 100 °C, the energy of thermal agitation is greater than the cohesive forces that bind the molecules together as a liquid, so water exists as a gas. The energy required to overcome the forces of cohesion is called the latent (or spent) heat of evaporation. It is called 'spent' because it cannot be used to perform useful work, being all used up in pushing the water molecules apart.

Below 100 °C, pure water vapour becomes pure water. As the water cools down, the thermal agitation of the molecules becomes less, with the result that more molecules pack into less space so the water increases in density, that is, a given volume of it becomes heavier. Liquid water is in fact structured—the molecules in it are held together in a more or less crystal-like lattice.

The next most important property of this remarkable substance is that at 4 °C it starts to turn from a liquid into a solid called ice (although the temperature at which it actually forms ice is 0 °C). Because solid water is a much more ordered crystal than liquid water, the water molecules are pushed apart during freezing. Thus water is at its densest at 4 °C and ice, being less dense than the water from which it was formed, therefore floats. If ice were heavier than water, it would sink, the oceans would freeze from the bottom upwards, and life as we know it could not exist.

Water is therefore present on today's earth as a liquid, a solid, and a vapour. But there was a time when this was not so. It was only when the temperature of our evolving planet fell below 100 °C that liquid water came into existence, falling under the force of gravity to fill the lowest lying areas.

As the earth began to solidify from a ball of hot gas and dust it became successively smaller and denser, and as its density increased so too did the force of gravity acting at its surface. Any substance free at the surface that was light enough to escape the developing pull of gravity was lost to space. One such substance was the rare gas neon which almost totally disappeared from the face of the earth. Some of our near neighbours in space, which in all probability were split off from the same ball of gas, have atmospheres that are comparatively rich in this

Solid water.

inert gas, because hand in hand with their greater masses go stronger gravitational fields that can 'hold down' the light gases.

This raises the question of how the earth managed to retain its water, for the water molecule is lighter than neon and therefore should have been lost to space in the same way. However the key difference is that neon is an inert gas. This means that it does not enter into chemical combination with other elements and therefore always has a free existence. In comparison, water is an extremely reactive substance, combining with many other chemicals and thus 'anchoring' itself to the crust of the solidifying earth. Thus the free neon disappeared to space, and the bound water remained on board this space ship earth.

Only later, when the crust had solidified and the pull of gravity was sufficient to maintain a wet blanketing atmosphere, could the water released from chemical combination have its own free existence on the surface of the planet in the form of the oceans. In contrast, the moon and Mercury are too small to retain any free water, and Mars with its larger mass retains only a very little, but sufficient to account for the exciting ideas of science fiction's little green men.

Once free water existed, the relentless cycle of evaporation and precipitation was set in motion. Thereafter the face of the earth was never to be the same again. Erosion not only wore away and smoothed the original rock surfaces beyond recognition, but the eroded material was carried as sediment and deposited elsewhere, covering other parts of the crust. Also, as the climatic pattern gradually became established, the temperature in some places fell below the freezing point of water and the process of weathering and hence of sediment formation was speeded up by the cycle of freeze and thaw brought about by diurnal and annual weather fluctuations. It is therefore impossible on any part of the earth's surface to find a record of the earliest phases of geological history, for the slate has been swept clean, the record erased by the action of water.

Recent study of the crust has however unearthed some remarkable facts concerning its formation. All the accruing evidence spells out very clearly that the crust of which the continents are built is very different from the crust

opposite
Common jellyfish, not so solid water.

Killer whale, 75% water.

that underlies the oceans. For a start the continental crust is geologically much older than the ocean floor. The mountain ranges of the continents are formed primarily of sedimentary rocks that have been uplifted and warped to form the mighty backbones of the land masses. The underwater ranges are much simpler. They are mountains of basalts and lavas poured out from cracks in the ocean floor, and are the direct result of massive volcanic activity.

Apart from their differing age and origin, the seamounts possess some unique features, not the least being that some of them have flat tops. For a long time it was thought that these flat-topped guyots had started life as ordinary mountain peaks which were then worn down by wave action. The only problem is that many of these flat peaks are more than 1000 metres below the contemporary sea level, a depth where wave action would never be felt, that is, unless the

A modern volcano, rock in the making.

oceans had been much shallower in the past. However all evidence indicates that although there have been minor fluctuations of the sea level, at a maximum these have been no more than 200 metres (even at the height of the ice age when an enormous amount of water was locked up in the solid ice sheets). The surprising fact is that the volume of the oceans has remained fairly constant since their origin. Another explanation of the flat-topped guyots was that the floor of the sea has subsided thus bringing them down to their present protected level. Unfortunately there are guyots to be found in areas where there is no indication of such subsidence, so how were they formed?

Proof of their genesis has come to light recently and it has come from a very unexpected place called the Afar Triangle, which is one of the hottest, driest parts of the earth. The Afar Triangle lies in Ethiopia on the south-west shore of the Red Sea. Here is a wild desert much of which lies below sea level, a rockscape pockmarked with volcanic ash and ribbons of fresh black basalt. It is a young landscape, as young as the ocean floor, and this is no mere coincidence for the Afar Triangle is actually a part of the floor of the Red Sea that has been cut off from the contemporary oceans. Immediate visual proof of this comes from the deep deposits of evaporites covering part of the basin to depths of more than 1000 metres. These are all that is left of the land-locked sea; the water has gone, evaporated by the fierce Ethiopian sun, leaving the salts behind.

The bedrock of the triangle is all of recent volcanic origin and a number of the volcanoes present in the area are flat topped, being composed of shards of volcanic glass. These are typically formed in underwater explosions due to the rapid cooling when the molten glass is spewed up into the cold water. There is little doubt that these now dry land guyots were formed underwater and that their oceanic counterparts are of similar volcanic origin.

One other remarkable feature of the Afar Triangle is that it has a central backbone of mountains much like the mid-ocean ridge that typifies all true oceans. Furthermore the central trough of the triangle contains no sialic (that is, continental) rock at all, for it is a continuation of the central axis of the Red Sea. Although it is now hot dry land, the ocean-making process is continuing. The trough is a central fissure through which new rock is being formed, flowing up from the molten magma below the crust. As new rock wells up and solidifies it is pushing the two sides of the Afar Triangle further apart, widening the ocean basin. It is now known that each of the main oceans has such a central trough running down the length of the mid-ocean ridge into which new rock is being injected. Successive outpourings of lavas plug the hole and in so doing widen the ocean by pushing the continents further and further apart.

Evidence for this fantastic ocean-forming process comes from another equally fantastic natural phenomenon. Throughout geological time the positions of the north and south poles have changed; in fact their positions have reversed on no less than one hundred occasions over the last seventy million years. Thus as the successive outpourings of new rock from the mid-ocean ridge have solidified they will have become magnetised in the direction prevailing at that time. If this theory of ocean genesis is correct, the ocean floor should be a complete magnetic record of the past taped permanently in the rocks. Extensive and exhaustive study has shown beyond doubt that this is true; the magnetic reversals are there for any magnetometer to read. The same tape record is there in the Afar Triangle and it is continued northwards along the length of the Red Sea—absolute proof that Africa and Arabia are being pushed further and further apart.

From all over the watery world evidence is building up to substantiate the hypothesis that the continents were once united as one large land mass. Then, about 150 million years ago, not all that long geologically speaking, this one super-continent began to split up, each part going its own way and the oceans as we know them forming in between. This is known as the continental drift theory.

The most surprising fact in this whole story is the constancy of the volume of oceanic water.

The sea has had its ups and downs but has never covered a considerably greater area of the land than it does today. Of all the questions still to be answered in this sphere of endeavour the mechanism of this equilibrium is among the most elusive. Along with the new rock welling up in the centre of our oceans comes new water. If this happens then there must be a roughly equal loss of water dissociated into oxygen and hydrogen in the upper atmosphere. The problem lies in obtaining proof.

Thus this planet was baptised and the incessant cycle that now typifies its surface was started for the first time—water, heated by the fire of the sun, evaporates to produce a vapour that remains in the air for a certain time before condensing and returning to earth. Earth, fire, air, and water—these were the four elements of the early philosophers who set out to understand the mysteries of this planet.

One such natural philosopher was Isaac Newton. Of the famous apple that fell on his head and awakened him to the reality of the force of gravity, 70 per cent was made of water. The rest of the apple, and for that matter of Newton's head, was made of simple geochemicals, that is, components of the earth's crust and atmosphere borrowed temporarily from the environment and compounded into the chemistry of life.

overleaf
Depths of the world's oceans. Some of the deepest parts are in the Western Pacific Ocean—for example the Philippine Trench (E of the Philippines) at 10,537 metres, the Kuril Trench (NE of Japan) 10,380 metres, the Japan Trench (SE of Japan) 10,371 metres, and the Marianas Trench (off Guam) 10,860 metres.

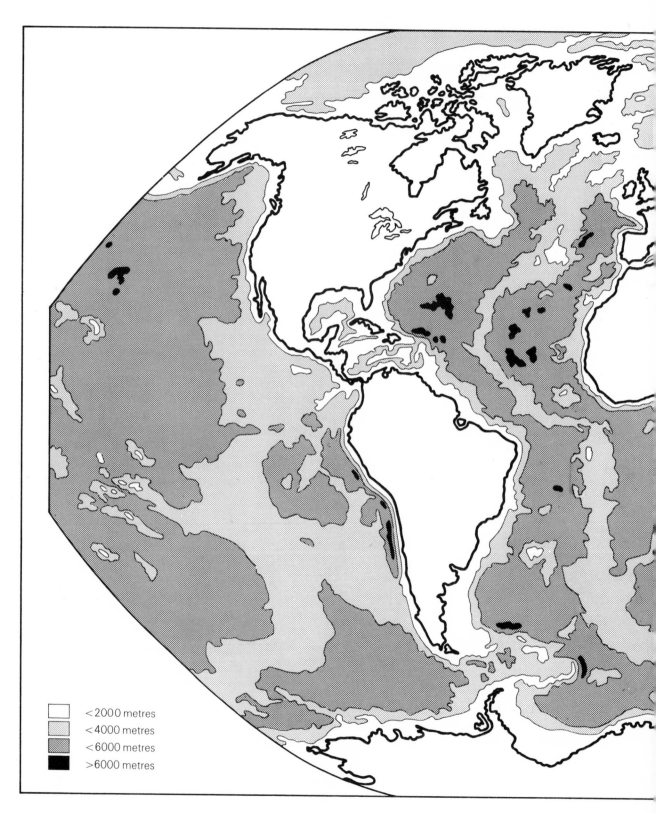

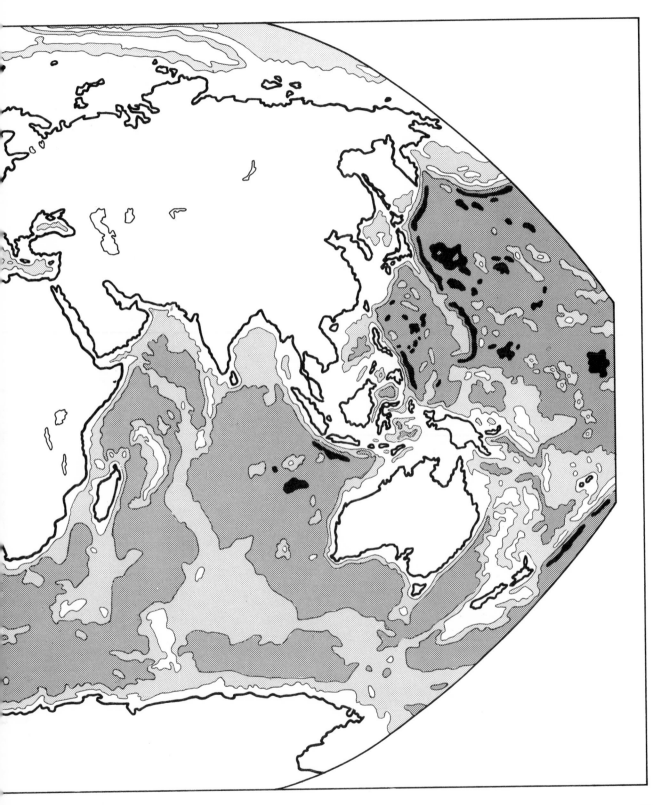

Environments in the sea

The primary source of the energy of life is the sun. The enormous amount of radiant energy that reaches the earth's atmosphere, almost unchanged after having travelled its eight minute journey through the nothingness of space, has to pass through a series of screening filters, the largest of which is the all-enveloping atmosphere itself. The atmosphere is a 300 kilometre thick blanket that consists of a mixture of gases in the following proportions: nitrogen 78%; oxygen 21%; carbon dioxide 0.03%; water vapour 0.01%; and lesser amounts of a host of other gases. Held in temporary suspension are numerous particles of inorganic dust and organic matter such as pollen grains and the spores of many of the lower forms of plant life, especially fungi and bacteria.

It is easy for most of us to understand the action of the atmosphere as a barrier shutting off light, for when a rain cloud comes along it gets darker, and we usually say that the sun has 'gone in'. It is perhaps more difficult to understand how the atmosphere can act as a filter that screens out certain parts of the incident radiation more effectively than the clouds.

The total radiation falling on the earth's atmosphere is made up of a range or spectrum of different wavelengths of energy, the whole being called the electromagnetic spectrum. Ultraviolet light, which lies at one end of this spectrum and which can seriously damage living organisms if received in too large a dose, is filtered out by dust and especially by water vapour. All ski and mountaineering addicts know only too well that, in the thinner atmosphere of high mountains, sunburn caused by too much ultraviolet can be very hazardous. Lower

down in the plains the danger is less because much more of the ultraviolet radiation has been screened out. At the other end of the spectrum is infrared, the part of the incident radiation that helps to warm the surface of the earth. Water vapour and carbon dioxide absorb the infrared radiation as it passes through, and they create a warm atmospheric blanket that acts as a gigantic greenhouse, helping to maintain the temperatures necessary for the living process.

We cannot see infrared radiation for the simple reason that our eyes themselves act as a sort of 'filter'. Vision depends on a chemical process activated by light, in other words, a photochemical process. This involves a pigment (visual purple) that undergoes chemical change due to the absorption of the energy contained in a certain portion of the electromagnetic spectrum. The part of the spectrum that brings about this change is naturally enough called visible light, and it contains all the colours of the rainbow.

Some of the visible light striking the surface of the ocean is reflected but some passes down into the water, and as it penetrates it changes both in quality and quantity. The water acts as a further filter and gradually removes the various component wavelengths, starting at the red end of the spectrum. Therefore the deeper you go the greater the predominance of blue light. The colour of the watery world below about 30 metres is a sombre mixture of blues, greys, and blacks, and nothing else (see page 218).

On the basis of light alone the oceans can be divided into three environments (see diagram on page 249). First is the euphotic or truly lighted zone which, because of differential absorption, is itself zoned both in respect to the quality and intensity of the light. Although it can have a maximum depth of 200 metres, this may be reduced to as little as a few metres in areas where the water is very turbid due to matter held in suspension. Then comes the bathypelagic zone, the twilight world of little light penetration. Its maximum depth is 1000 metres. Below the twilight world is the great volume of total darkness called the abyss. In its eternal blackout nothing can ever be seen unless the particular object produces its own light. This is the great inner space of our own planet about which we have much to learn.

In the same way the oceans are zoned in relation to light, they are zoned in relation to temperature. The surface of the sea is warmed by the radiant (infrared) heat of the sun during the day, and the heat is stored in the great mass of water. Each night part of this stored heat is re-radiated back to the cooler atmosphere above. In the tropics, where the daily cycle of radiant

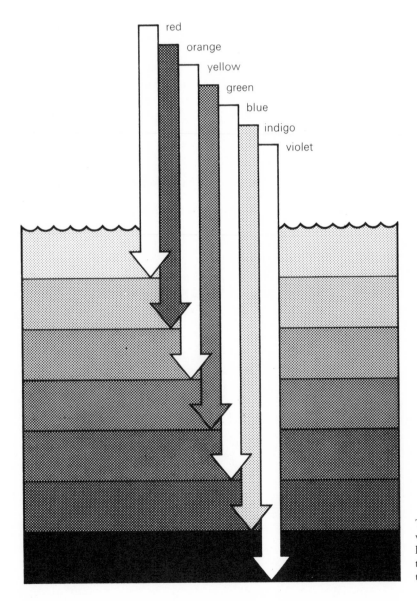

The water column acts as a filter on visible light. The depth to which light penetrates varies according to the position of the sun and the turbidity of the water.

energy is always approximately twelve hours day and twelve hours night, the surface temperature rarely falls below 30 °C. Nearer the poles where winter nights are long, sufficient heat is lost from the great storage heater that the surface temperature drops below 0 °C and the sea freezes. If it were not for the remarkable property of water already described (that it is densest at 4 °C), the ice would sink; instead the dense water at 4 °C settles down into the abyss, which thus becomes a giant cold store maintained at this regular temperature.

There is one other temperature environment, and that is the zone between the waters of the surface and the cooler waters below. This zone, which may be quite abrupt, is called the thermocline or the zone of thermal change. Its exact position in the sea depends on all the variables affecting the surface temperature. If you are diving in temperate waters it is not an unusual experience to swim from the warm waters above through a thermocline into the cooler waters below—the transition experienced in this way can come as quite a shock.

For a long time the only way to find the position of the thermocline was to take the temperature of the water. However quite recently John Woods, meteorologist and diver, developed a novel method of location by using coloured dyes. His answer is simple—dye released from small packets congregates along the thermocline—but the results show that they are anything but simple. There may be several and some may change their position quite dramatically. However thanks to this new method we can now locate them more easily and can start studying them in great detail.

Thus, as far as energy is concerned, the four most important environments under the sea are: the euphotic zone with its warm, well-lit waters that are subject to large fluctuations in light and temperature; the abyss, the largest part of the great oceans, a zone of total darkness and great thermal stability; between them is the twilight zone where a certain amount of light penetrates, but not really enough to affect the environment; and the thermocline, a complex zone of thermal change.

The surface of the sea

Whether it is the gentle ripples stirred into the surface of a pond by a falling stone, the incessant ebb and flow on a sandy beach, or the awesome rush of water onto a rocky shore at the height of a storm, waves hold a certain fascination for most people, especially when viewed from a safe perch on dry land. At sea, from the pitching deck of a small boat they are perhaps a different matter, although just as intriguing to watch.

There are a number of different sorts of waves, but the ones most commonly seen are caused by the wind. Wherever the wind blows across the sea, waves will be created—try it out in your bath by gently blowing over the surface of the water. The area that is directly affected by the 'blow' is called the generation area: in the bath it may only be a few square centimetres; in the open sea it may be more than 2500 square kilometres in extent. In this generation area, the wind loses some of its energy as frictional drag that ruffles the surface of the water into a series of waves. The smallest will immediately break, losing their energy in a turbulence of white water. The larger the wave the more easily the wind can add more energy to the crest. Short ripples are thus built up into long waves that can accept more and more energy and can grow in height without breaking. It is in this way that a whole series of waves, or wave train, can be built up producing the regular rolling motion that sets up such qualms in the stomach of a landlubber.

It is very difficult to describe the waves within the generation area, for there are so many of them of different shapes, heights, and lengths. Only when a regular train has been set in motion can their properties be predicted and their energy easily determined. Such waves have length,

measured from the top of one crest to the top of the next; height, measured from the base of a trough to the top of the next crest; and period, the time taken for two successive crests to pass a given point. From this information it is possible to calculate the energy of a wave, a very important feature for anyone who has to deal with them. The only problem is that when knowledge of the force of a wave really matters they are not usually in a nice, orderly train and there is no time to stop and measure them. The old tale that every seventh wave is a big one is, sadly, only a tale, and one with no foundation at all, although if you sit and watch them for long enough it is surprising how easy it is to believe it.

Waves breaking against a vertical rock.

There is however one rule that waves do appear to obey: over a long fetch (the length of the generation area), long waves will predominate, and it is under these conditions that the really big ones will be formed. Also, generally speaking, the longer the fetch, the higher will be the waves. It is difficult to say what the record is for a wind-generated wave, because many of the guesstimations can be relegated to the category of worried fishermen's tales. The real problem is that the big ones occur in the open ocean and it is very hard to estimate height from a platform that itself is moving up and down on the thing you are trying to measure. However in 1933 the USS *Ramapo* recorded a truly mountainous sea with at least one wave that afterwards was calculated to be over 34 metres high. The ship was running before a wind that had gradually increased to a speed of 60 knots over a period of several days, indicating a very long fetch and accounting for the record wave.

This is of course not the largest wave on record, simply the largest wind-generated wave, the height of which was scientifically calculated. The real monsters, the so-called tidal waves, are generated neither by the wind nor the tide, and they perhaps should not be called waves at all. Tidal waves, or to give them their correct name, *tsunamis* (Japanese for tidal wave), are generated by sub-marine upheavals such as earthquakes, volcanic explosions, and landslides. These massive movements of the earth cause massive movements of the oceans, and that can spell real trouble for anything living at sea level in the near vicinity.

The most infamous *tsunami* was that associated with the volcano Krakatoa in Indonesia. Krakatoa exploded and virtually ceased to exist on 27 August 1883, with the result that at least three cubic kilometres of rock shot up into the air and landed in the sea. This caused an immense displacement of water which, together with the actual force of the explosion and the accompanying less spectacular earth movements, was enough to start a shock wave of incredible dimensions, whose effect was detectable more than 18,000 kilometres away. Great waves raced out across the ocean at speeds in excess of 640 kph and left about 36,000 people dead in their wake. Fortunately big ones like this are rare, but even the effect of the small ones occurring each year cannot be ignored.

The more normal wind-generated variety can also be enormously destructive out at sea, especially if they catch a ship in the charge of an unable seaman. However waves wreak their main havoc inshore when their energy is rapidly dissipated against a non-mobile object. To understand this it is necessary to know exactly what it is that makes a wave tick.

In a wave the water particles in the surface layer are not moving forward: they are in fact going round and round, the diameter of their orbit being equal to the height of the wave. Lower down in the water mass the particles describe smaller and smaller circles. Thus the water body does not itself move forward—the energy of the wave simply passes through it with only the top few metres being affected at all. In this way the energy of the wind is 'stirred'

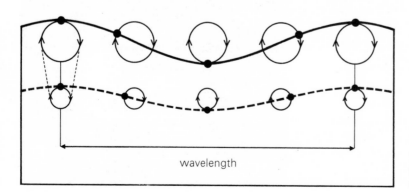

wavelength

The orbits indicate the movement of the water particles in a wave; the deeper down the less the movement.

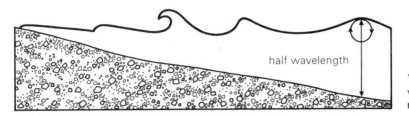

half wavelength

When the depth of water beneath a wave equals half the wavelength, things start to happen.

into the water mass in the generation centre. Once outside the generation centre the wave pattern settles down to the rolling swell that typifies the open ocean.

The forward progression of the wave front will continue unheeded until it reaches shallow water. The critical point comes when the water depth equals half the distance between the waves. As the solid bottom begins to affect the speed of the wave front and the character of the waves, the whole pattern is changed—the even swell is about to be transformed into the excitement of a breaker. The effect of the shallow water is to turn the wave front so that it runs parallel with the shore. As this happens (and it is a very easy thing to observe for yourself), the waves increase in height and the circular orbitals are tipped over, producing the line of breaking water. Waves approaching a gradually shelving sandy shore, like the famed Bondi Beach, produce the enormous curling rollers that are such a delight to expert surf riders. However, where

waves with a similar amount of energy impinge on rapidly shelving rocky shores, the effect is anything but delightful. The waves crash over, entrapping a large volume of air that is released under pressure, sending fountains of spray skywards in great explosive waterspouts. The force of such waves dissipated abruptly against vertical cliffs is sufficient to hurl large rocks over a hundred metres into the air, depositing them on cliff tops and smashing the lanterns of ill-protected lighthouses.

Whatever the strength of the breaker, be it the spectacular dash of the storm or the more constant ebb and flow of lesser waves, the effect is the same—the incessant erosion of the shore rock down to boulders, pebbles, sand, and silt; a gradual reduction in size until the geochemicals that once made up the solid rock disappear as a puff of minerals dissolved in the sea.

A quartet off Bondi Beach, Australia. Depth is less than half the wavelength.

Salts of the sea and the cycle of energy

For a long time, men who studied the chemistry of the sea (and there is a special name for them, thalassochemists, derived from *thalassa*, Greek for the sea) thought that the primeval oceans consisted of fresh water. They argued that in the incessant cycle of evaporation and condensation —sea, to atmosphere, to land, to rivers, and thence back to the sea (the hydrological cycle)— the erosion of the dry earth gradually enriched the waters, in time producing the salts of the sea.

Fired with this idea they also argued that, knowing the total quantity of salt present in the oceans and the rate at which it was being added by the rivers, it should be possible to calculate a time scale for the existence of the oceans and hence of the planet. The answer they came up with was a mere 90 million years, a figure far short of most other estimates. Added to this is the fact that the proportions of certain of the dissolved salts are not consistent with those calculated in the knowledge of the abundance and solubility of the geochemicals in the surface of the crust. Furthermore, modern measurements have shown that the salinity of the sea has remained unchanged for at least 2000 million years. Therefore the source of the bulk of the salts cannot be simple erosion.

The salt mill under the sea.

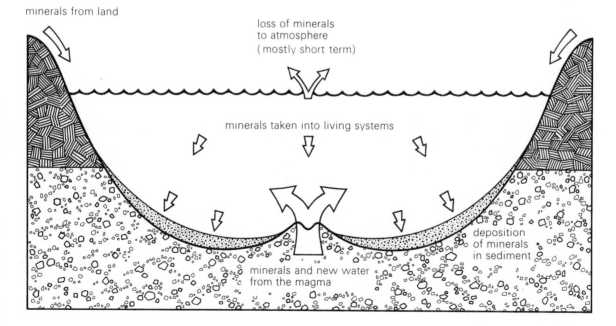

minerals from land

loss of minerals
to atmosphere
(mostly short term)

minerals taken into living systems

minerals and new water
from the magma

deposition
of minerals
in sediment

It is now known that the figure of 90 million years probably represents the residence time for sodium in the sea, that is, the actual time an average sodium atom remains dissolved in the ocean once it has been added. But where does it come from? and, perhaps more to the point, where does it go to?

There is an old Norse folk tale that attributes the saltiness of the sea to a magical mill that sits at the bottom of the ocean and continually churns out new salt. Ironically, as the studies of the thalassochemists are getting underway, they are coming to a similar conclusion.

The location of the mill is reckoned to be in the mid-ocean rifts separating the strong plates of the earth's crust that bear the weight of the continents. The crustal plates are moving apart, ever so slowly carrying the continents with them as they go their opposite ways causing fresh basalt to flow up from the molten mantle below to heal the wound. As the new rock is formed so new water carrying dissolved minerals is released and added to the volume of the oceans. The minerals include many heavy metals that rapidly come out of solution, and this is how one of the great natural treasures of the deep is formed—the mysterious manganese nodules that litter the ocean floor.

Here then is a source of at least some of the salts. Once added, they are of course subject to modification by factors other than manganese nodulation: additions from volcanic activity, and depletion as certain minerals are locked up under the enormous pressures of the great deep into more extensive sediments, or are entrained into the web of life close to the surface.

The salts of the sea are thus best regarded as part of a very long term cycle which we are just beginning to study at one very short point in its history. A cycle it may be, but the concentration and proportions of the elements are surprisingly constant, so much so that it is possible to make a complete analysis of sea water just by measuring the amount of chloride present. If this is done with great accuracy then from our knowledge of the constancy of make-up we can calculate the concentration of all the many other dissolved elements.

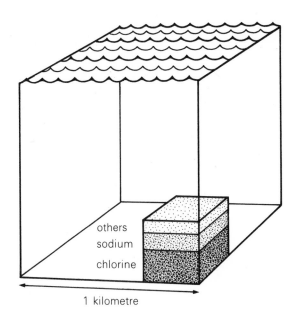

One cubic kilometre of sea water contains one million million kilograms (1000 million tonnes) of water, a lot of sodium and chlorine, and many other things, including 3500 kilograms of arsenic and 5 kilograms of gold.

There is one other way in which salts are removed from the sea that at first sight is most surprising, and that is the continual loss to the atmosphere (although much of this is very short term). As the wind stirs up the ocean into those breaking waves, tiny droplets of water are whipped up and evaporate to form water vapour. The minute salt particles left are small enough to be carried upwards in the moving air mass. When they reach a height where the atmospheric conditions change, the salt particles can act as nuclei for the recondensation of the water, which then falls as rain either directly into the sea or onto dry land, eventually to return to the oceans via the rivers. In this way salts are continually being recycled but, much more important, energy is being transferred at the same time.

Evaporation requires energy and that energy, called the latent heat of evaporation, comes from the surface of the sea which is itself cooled in the process. At higher altitudes, under reduced pressure and especially where the tiny salt particles are present, condensation releases the energy back to the atmosphere. The water

Cumulus clouds, water and a little salt.

Cirrus clouds, high altitude water.

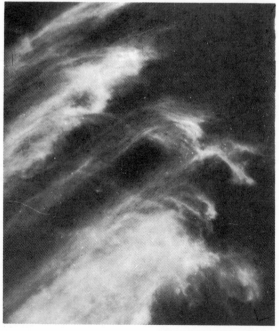

falls as rain and the atmosphere is left just a little warmer. Latent heat of evaporation is thus transferred from the sea to the atmosphere, with the net result that the atmosphere is warmed by the sea.

This is of course not the only source of heat, for the sun does its fair share of adding energy each day. Thus the overall pattern is made much more complex. Also, because air is thinner than water, the two experience differential heating: the total capacity of the atmosphere to hold heat (its thermal capacity) is equal to that of only the top three metres of ocean water. As heat is lost from the surface waters to the air, the cooled water becomes more dense and sinks down through the thermocline until it joins the permanently cold body of water at 4 °C that makes up the great volume of the ocean deeps. There it remains, only slowly being recycled to the top.

The surface wedge of oceanic water is thus a gigantic storage heater that helps to ameliorate

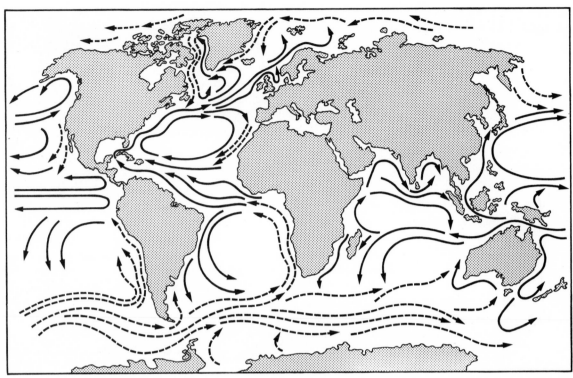

- - - - - - -► cold currents
━━━━━━━━━━► warm currents

The circulation of the sea.

the climate of adjacent land masses. And this brings in another complicating effect. Because of the heat capacity of the oceans, the waters of the earth heat up and cool down much more slowly than the adjacent land masses. It is for this reason that Winnipeg, for example, which is situated near the centre of the continental land mass of Canada, suffers a climate of extreme temperatures, with a mean summer maximum of 26·7 °C and a mean winter minimum of − 22·3 °C. Compare these with the figures for the town of Victoria, situated on the Pacific coast at approximately the same latitude: 20 °C in the summer, but only 2 °C in winter. This is the true 'power' of the sea.

It is this differential heating, together with the change of the seasons between the northern and southern hemispheres, that causes the winds to blow—the same winds that whip up the waves (exactly where this chapter came in), producing extensive cloud masses that shadow large areas

of the earth, in turn creating further differential heating.

So it goes on, the interaction between the land and the sea. The hydrosphere and atmosphere are so complex that it is indeed surprising that there is any pattern either in the winds that blow or the currents that flow that may in the long term be relied upon. The fact is however that there are, and they can: the arrows that spiral their way across the pages of even the simplest schoolboy atlas are real, measurable phenomena, marking the major patterns of the complex interplay of the forces that keep the sea and the atmosphere in constant motion.

Perhaps then it is to be expected that at the meeting point where these three great moving systems, the air or atmosphere, the sea or hydrosphere, and the earth or lithosphere, meet and mix that a unique happening should have taken place. That happening was the evolution of life—the very subject of this book.

33

The first flicker of life

At the edge of the sea, in the mix of earth, air, and water, especially where the fire of the sun evaporated the water causing concentration of the salts in shallow pools, a brand-new chemistry was to develop. Of all the elements present in this environmental cocktail only four were to play a backbone role in the new chemistry called life. These four were oxygen and hydrogen (the constituents of water), and nitrogen and carbon (both of which were present in the atmosphere). All that was needed to bring about the transformation was a source of energy, thus wherever volcanoes poured their lavas into the sea, or lightning struck the earth, the vital synthesis was possible. So it was that the first life chemicals —carbohydrates, fats, and proteins—came into existence.

Another possible site of synthesis was in the great deeps where pressure and the magical salt mill could have provided the right conditions. Exactly where and how many times such syntheses occurred we shall never know, but it must be stressed that there is neither evidence to suggest nor reason to believe that it only happened once.

Thus over 3000 million years ago, when conditions were just right, dilute solutions of these new chemicals were first formed. They were in the strict sense organic, that is, of life, but they could not yet be regarded as living, and many further changes had to occur before it could be said that life had evolved. Modern research has found the frail imprint of the life chemicals in rocks that are older by far than those containing the first organised fossils.

The key development in the early stages of organisation was the evolution of a membrane that could contain and protect the new organic compounds. We can at present only surmise as to how it happened, but the evidence we see, both in the living world today and the record in the rocks, tells us that the first organisms consisted of envelopes made of simple fats and proteins—envelopes that could allow and yet control both the ingress of the raw materials for, and the egress of the waste products of, living. It was in this way that the pristine inorganic, non-living environment first became 'changed' (though with the record of recent times perhaps the word should be polluted) by the products of living organisms or, to be more correct, by the complex chemistry of life.

At this stage life was a very localised phenomenon and it would have remained so but for the next main step in evolution. The word organism simply means organised life chemicals, and just as energy was needed for their synthesis so too was energy necessary to maintain their organisation. The only energy source available to these first organisms was from the other life chemicals floating about in the pools around them and, however efficient and widespread the process of synthesis, they must have been in short supply.

So it was that life faced its first power crisis. If it had not been solved, either life would have remained at a very low ebb or, more probably, it would have disappeared altogether. The solution to the problem was the evolution of a new life chemical, a green pigment we now call chlorophyll. Chlorophyll is green because it has the property of absorbing all the wave-

The sea lettuce *Ulva lactuca* (see page 53).

lengths of visible light except the green that we see, which is reflected. The absorbed light was just what the evolutionary doctor ordered —an almost limitless source of energy beaming down from the sun which could now, thanks to chlorophyll, be channelled to organise and drive the evolutionary process.

The fascinating thing is that it was not only chlorophyll that was necessary; the pigment would have been useless without a whole complex of other life chemicals that together constitute the organised controllable pathways of photosynthesis. A complete description of the photosynthetic process could easily fill the rest of this book, and the more detail we considered the clearer it would be that we do not begin to understand how the complete system ever evolved.

Suffice it to say that it did. The components of chlorophyll are there in the rocks that pre-date the first real fossils. Furthermore the first true fossils closely resemble organisms inhabiting the seas today, absorbing and fixing the energy of the sun to produce and maintain new organisms. Such observations cannot be regarded as proof, but together they provide an impressive body of circumstantial evidence as to the hows and wherefores of the beginnings of life.

Of all the multitude of biological facts, there is one more intriguing than any other: by using all the highly sophisticated methods and machines of modern research, detailed studies have shown that the majority of all organisms alive today share the same basic chemistry of life. The cells that make up the human brain, those that support the branches of great tropical trees, and those that constitute the blades of grass around your feet, each have their own chemical peculiarities but their mechanisms for transforming energy into the organisation of living are all basically the same. These mechanisms must have developed in the dawn of evolution and are best regarded as the basic chemical information required to maintain and sustain the living state, and it would appear that they have changed but little throughout much of evolutionary time. (For a diagram of a simple cell, see page 40.)

If, as we believe, the first information of life was formed by a series of chance happenings (the chance of the right type of water being in the right place at the moment that the right amount of energy was added), then one important factor would have been the ability to conserve the vital information, for once lost, evolution would have had to wait for another series of chance happenings. So a further key to evolutionary success was the development of a chemical memory bank that could store the relevant information, passing it on from one generation to the next.

Long before modern science discovered the chemical nature of the data bank, it was known that part of the information resided in a specialised region of the cell called the nucleus. The discovery of the nature and the workings of the nuclear substance has filled, and is still filling, many of the leading scientific journals of the day. The men who made the key discoveries that opened up the story of life itself were Maurice Wilkins, James Watson, and Francis Crick. In 1962 they received the Nobel Prize for their work on the substance on which life is centred, namely deoxyribonucleic acid, which has, at least in its capitalised form of DNA, become a household word.

Carbohydrates, fats, and proteins, these form the stuff of life; chlorophyll is the transformer of light into living energy; and DNA is the substance with the long memory that is able to store and, with help, replicate the information of life. This was the package deal of chemical evolution, a do-it-itself kit that was to transform the non-living world into a living one.

Rhodymenia palmata or dulse, a red seaweed (see page 53).

37

Bags of chemicals

A bag of chemicals—not much with which to conquer the world—yet that is how life began. Today, a mere 3000 million years later, there are few parts of the world devoid of life, and still the most abundant types of organisms alive now are just simple bags of chemicals. Many of them are so simple that it is impossible to decide whether a particular type is a plant or an animal, so the best thing is to avoid making mistakes by using the all-embracing term, the *Protista*.

The simplest of the protists are the bacteria and blue-green algae that at first sight appear to lack any internal organisation. Because of this, they have been given the special name of prokaryotic organisms, a name meaning 'before the cell'.

Only with development of the high-powered light microscope and especially with the inven-

Hairy bags of chemicals—the hairs of the flagella that propel marine bacteria through the water (magnification ×8500).

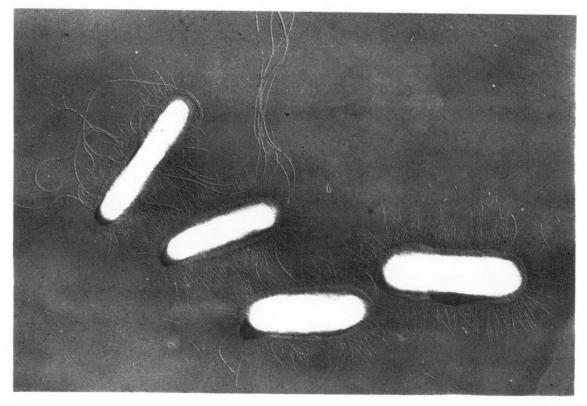

tion of the electron microscope did the internal organisation of the prokaryotic cells become apparent. The problem was one of resolution. If you have good eyesight you should be able to distinguish between two tiny objects 0.1 millimetres apart. If the two objects are closer together then the unaided human eye will see them as one, for it cannot resolve the difference. Perfection of the light microscope allowed two objects only 0.0002 millimetres apart to be resolved, but that only under perfect conditions with best blue light. The electron microscope dispenses with the limitations of light and uses a beam of electrons to solve the problem. The best electron microscope should resolve objects about 0.0000004 millimetres apart. That is just about four times the size of the world's smallest atom, hydrogen—surely good enough for anyone!

The optical microscope is thus about 500 times as powerful as the human eye, and the electron microscope approximately 500 times as powerful as that. Here power refers to the power of resolution, not magnification, for with the right set of lenses there is no limit to magnification. Under the resolving gaze of the electron microscope even the most complicated cells reveal their structural secrets: ordered arrays of macromolecules, DNA, chlorophyll, proteins, fats, and carbohydrates, each one fitted to play its own particular role in the living structure.

Simple as the bacteria and blue-green algae are, some of them perform a function without which evolution would have been severely handicapped from the very beginning. The problem was nitrogen, for although it constitutes 78 per cent of the volume of the earth's atmosphere, it is a relatively inert gas. Hydrogen burns, oxygen supports combustion, but nitrogen does very little, at least chemically speaking. Thus right from the start it was difficult for life to do anything with atmospheric nitrogen.

Evolution overcame the problem by furnishing certain organisms with a chemical pathway by which nitrogen could be fixed (removed from the atmosphere) and converted into reactive compounds called nitrates. Nitrates

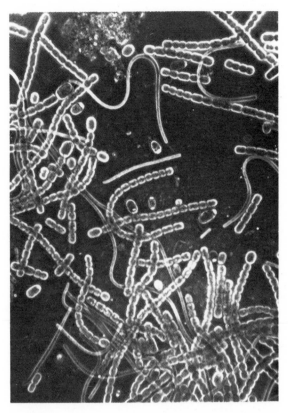

Anabaena, a blue-green algae: chains of simple cells (magnification × 480).

are all extremely soluble in water, which is perhaps another reason why life began in the sea for, as the earth cooled below 100 °C and liquid water fell onto the earth, all the nitrates present on the surface would have been rapidly dissolved and washed into the sea. The resultant dilute solution was probably sufficient to set the wheels of evolution in motion, but not to keep the evolutionary machinery supplied with this key raw material. Thus the capability of nitrogen fixation was built into the evolving complex, a capability that was and has remained unique to the prokaryotic organisms.

With time, the complexities of organisation increased and specific parts of the cell were set aside to perform specific functions. It was these functional sub-units or organelles that the early microscopists described in accurate detail: chloroplasts containing the all-important green pigment, the nucleus with its memory banks

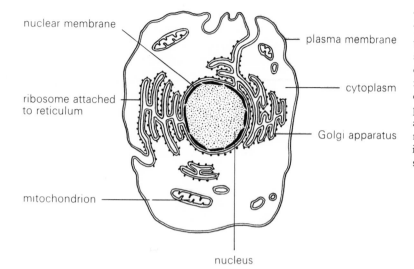

nuclear membrane

plasma membrane

ribosome attached
to reticulum

cytoplasm

Golgi apparatus

mitochondrion

nucleus

The living contents of a living cell. The plasma membrane is a double wrapper that allows raw materials in and waste products out; the nucleus contains the information of life; the mitochondria are the power houses of the cell; the ribosomes are the protein factories; and the Golgi apparatus is concentrations of double membranes that appear to play an important role in the production of secretions from the cell.

full of living information, vacuoles, tonoplast, mitochondria, Golgi bodies, and internal membranes. These were the first true cells.

All this was made possible by the craftsmanship of the Victorian microscope technicians—men like Ross, Powell, Leyland, Beck, Watson, and Collins, each of whom made their instruments with loving care. The crowned heads of Europe vied with one another for resolution and the optical system was pushed to its limits. The ultimate test of any good instrument was to resolve the fine markings on the skeleton of *Amphipleura pellucida,* which are exactly 0.00025 millimetres apart.

Amphipleura pellucida is a kind of microscopic plant called a diatom, and diatoms are without doubt among the most abundant and most beautiful plants to be found growing in the oceans. Their abundance relates to the fact that many of them are planktonic, riding high in the lighted waters of the euphotic column. Their beauty resides in their skeletons which are made of purest silica.

Although diatoms come in various shapes and sizes, they all revolve around one theme. Each skeleton, or to give it its proper name, frustule, consists of two halves that fit together like an old-fashioned pill box, the lid being slightly larger than the box itself. The living substance of the diatom is both on the inside

and the outside of the frustule, and many of the sculptured patterns that still so excite microscopists are in fact holes that go right through the frustule. (For the top view of a diatom called *Triceratium grande*, see endpapers; magnification x75.) Inside the box, safe from harm's way, are the bulk of the organelles, for diatoms are true cells. They are however true cells with a problem, and the root of their problem lies in their beautiful skeletons.

The basic method of reproduction of all protists is simple division into two equal halves. This is of necessity preluded by a division of the nucleus and a replication of the information it contains. Once this has been completed the cell splits into two halves each of which in time grows to the size of the original parent cell. But not so with the diatoms: one daughter cell inherits the top of the pill box, the other the smaller bottom half. And here is the rub, because each daughter cell then sets about making a new bottom half. This is all very well for the first couple of divisions but, as the family tree grows, one line of daughter cells stays the same size while the other gets progressively smaller and smaller. If allowed to continue, the end result would be a cell so small that it could no longer contain all the chemical information necessary to keep it ticking. However, ·long before this point is reached, division stops in

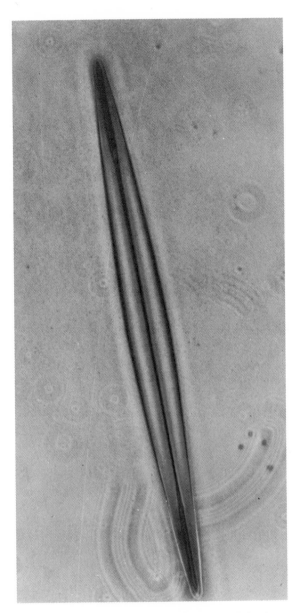

Amphipleura pellucida, a test for resolution. Each fine line is 0.00025 millimetres apart (magnification × 440).

Victorian as the drawing room in which they were gathered. Many of the specimens they observed were dead but the invention of the live box, an attachment just like a miniature aquarium, made possible investigations of living protists: how they moved, how they reacted towards light, and how they reproduced, multiplying their delicate forms. Although the diatoms were the star performers, two other groups of protists held the attention of the microscopists both for their beauty and for the problem of resolving their parts even with the best instruments.

One of these was the dinoflagellates which unlike the diatoms have a skeleton of cellulose, and what is more they wear it on the outside.

A pile of pill boxes: *Fragilaria,* a colonial diatom (magnification × 540).

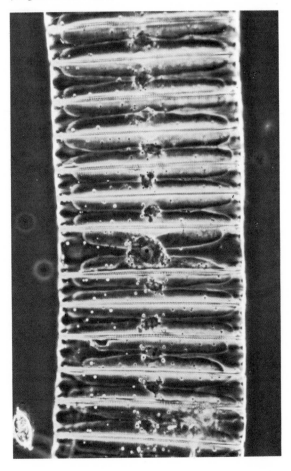

the diminishing line and the tiny cells undergo a period of rest during which they grow back to the size of great great great . . . grandmother.

When the master of a wealthy Victorian household unlocked the microscope box on a Saturday evening, the family would gather around to take their turn to look into the world of the protists, a world that was as ornately

41

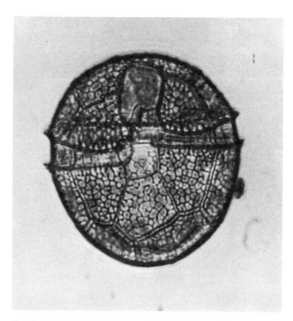

Peridinium, a dinoflagellate (magnification × 1120).

They are also much more mobile than the diatoms actively swimming through the water propelled by the rhythmic lashing of two whip-like organelles, the flagella, which protrude out through the armour plate: hence the name of the group, dino (two) flagellates (swimming organelles).

The 'dinos' are certainly one group of organisms that has not yet managed to sort itself out, animal from plant-wise that is. An outer coat of cellulose says 'plant' yet their capability of actually swimming about says 'animal'. Some members uphold the plant side by containing chlorophyll and fixing the energy of the sun. Others lack this pigment and actively feed on smaller organisms. And, just to confuse the picture even more, some have the capability of feeding like animals in the dark and of photosynthesising when the sun is shining.

Perhaps this adaptability is part of their key to success, for of all the protists in the sea, these are undoubtedly among the most successful. These tiny 'plant-animals' are found in every ocean of the world and in an average sample of open ocean water there can be as many as 300,000 per cubic metre. Such facts immediately make one realise just how small these organisms are. Next time you are at the seaside fill a glass bottle with sea water and take a close look. Once the sand and silt has settled out it will look like pure sea water, yet bottled up in there are tens of thousands of living cells. One interesting experiment is to cover the bottom of the bottle with tin foil and then leave it standing in a well-lit spot. If you are lucky, after a time it will be possible to see that the water above the tin foil is slightly cloudy or even faintly coloured. The reason is that many of the dinoflagellates possess light-sensitive organelles and actually swim from the dark to the lit end of the bottle.

However don't be disappointed if it does not work the first time; the golden rule in all experiments is to try again. I well remember my excitement when, having forgotten my experiment until early in the evening, I found that the top half of the bottle was glowing with an eerie light. Luck had been with me, for the water sample contained the large, naked dinoflagellate with the very apt name of *Noctiluca scintillans.*

Many planktonic organisms have the property of producing light. The reason is obscure but the effect can be very startling, as two bathers found when they went swimming one night near a new atomic power station on the British coast. They came out glowing from head to foot, not as they imagined because of radioactivity, but because of a phosphorescent dinoflagellate that just happened to be very abundant in the coastal waters at that time.

In some parts of the world, short-lived superabundances of certain types of dinoflagellates are a common although poorly understood phenomenon. Such planktonic population explosions are called blooms and they may have disastrous side effects. One regular bloomer is *Goniaulux,* which has the added distinction of being red, and when it blooms the sea can turn

opposite
Plankton, minute plants and animals that float near the meat and vegetable soup of the sea (magnification × 360).

42

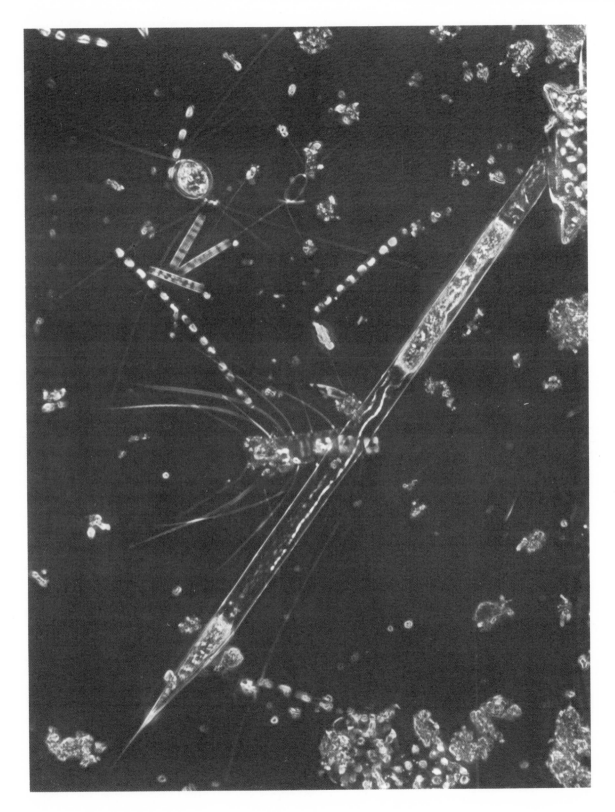

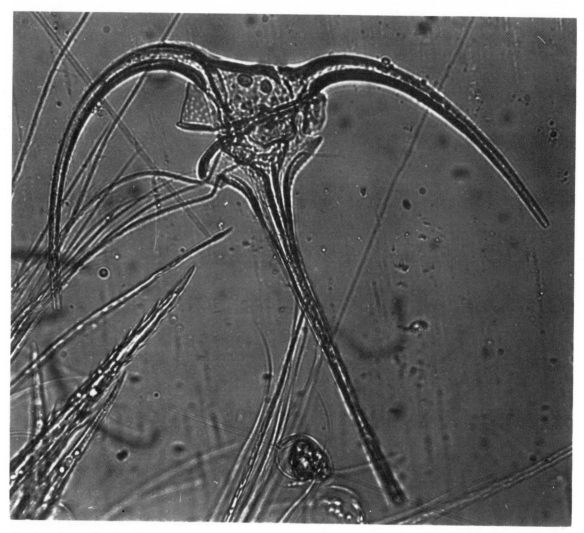

Ceratium tripos, a dinoflagellate. The long 'horns' increase the surface area, thus aiding floatation (magnification × 175).

the colour of tomato soup as their numbers soar into tens of millions per cubic metre. Apart from the red pigment, *Goniaulux* also produces a substance which, although harmless to itself, is a very toxic poison, attacking the nervous system of more highly developed organisms like man.

As we do not directly feed on plankton, and certainly never drink enough water to obtain a fatal dose, there would seem but little chance of us ever being affected by the poison.

Unfortunately certain shellfish, such as mussels and oysters, feed by straining succulent plankton out of the sea water, so they can get much more than their fair share of the poison, but being simple animals they are unaffected by it. However a quick dozen mussels as an *hors-d'oeuvre* and the gourmet can end up in deep trouble. For this reason along the coasts of the world where *Goniaulux* blooms are a common occurrence the collection of shellfish for human consumption is banned throughout the 'red tide' season.

If red water means trouble, then white water means just the opposite, at least in the parlance of the herring fishermen of the North Sea. White water is caused by the superabundance of members of another group of armoured protists that go under the lovely name of coccolithophores. The milkiness is due to their armour plates being made of chalk; this strongly reflects the light so that it appears as if the surface of the sea is lit from below. The first evidence of the existence of these tiny plants was the continual discovery of their ornate chalky scales dredged up in the ooze that covers the bottom of the oceans. These peculiar structures remained a great mystery to microscopists until it was realised what they were. The herring

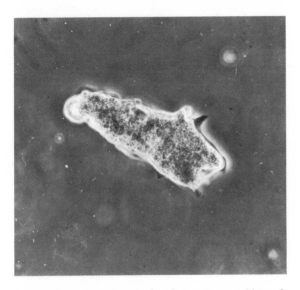

An amoeba, not much to look at but an immortal bag of chemicals.

fishermen reckon that white water means that there are a lot of herring about, and there may well be something in that supposition. These chalky creatures are in actual fact much more abundant in the warm seas near the tropics than in the frigid northern waters, so white water in the North Sea could indicate a large input of potential fish food brought north in the warm currents from the tropics.

To conclude this very brief dip into the world of the protists, one group of organisms must be mentioned because more than any other they come closest to the title of this chapter. They are the amoebae. One of the larger amoebae goes under the delightful name *Chaos chaos*. The only order in the chaos appears to be a nucleus and a granular ground substance. It has neither a back nor a front end and when it moves it does so by sticking part of its anatomy in the direction of go and the rest flows along behind. *Chaos chaos* feeds by ingesting other organic matter, and it can sense the environments through which it moves, and respond to stimuli both good and bad. By means of a special organelle, the contractile vacuole, it regulates its own internal environment, and when fully grown it reproduces by dividing into two equal halves, each of which looks just like mum.

There is no getting away from the fact that this simple cell is an animal and that biologically speaking it does just about everything that all other living things do. Much more than that, the living bag of chemicals is potentially immortal. At each division, half of the living substance, or cytoplasm, is shared between the two daughter cells. If evolution started off with an organism like *Chaos* then its component molecules could well be around in its direct descendants cavorting about in the sea today.

From here on up (or is it down?) evolution went from ordered chaos to specialisation, and with specialisation the potential of immortality disappeared.

At the edge of the ocean

It is a strange fact, but borders hold a special interest for many people and none more so than the one that separates the land from the sea. The biologist's preoccupation with the edge of the ocean probably stems from the rhythmic pattern of environmental change caused by the ebb and flow of the tides.

Tides are massive rhythmical waves generated by the forces of gravity that are exerted on the earth by its neighbours in space—especially by the moon, our only natural satellite, and the sun, our nearest star. All the other planets of the solar system and all the other stars of our galaxy also exert a gravitational influence on the earth but their effect is swamped by that of the moon (which is very close) and the sun (which is both close and very large).

As the great mass of the moon travels around the earth it attracts the waters of the ocean to the extent that a great hump of water is formed that tends to remain exactly under the moon as it travels on its orbit. At the same time, a similar, compensating hump or wave crest travels around the earth on the opposite side. The period of this true tidal wave is about half a lunar day and the effect we see from our safe

opposite
Patterns in the sand at the edge of the sea.

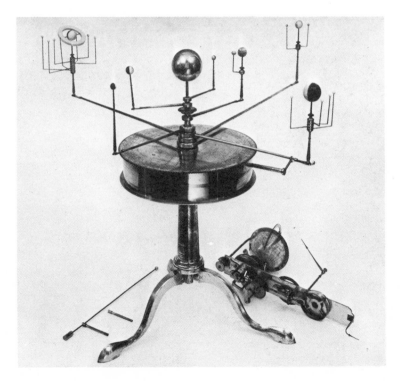

A late eighteenth century orrery, a working model of the solar system. It shows, among other things, how the tides are generated.

47

perch on land is that the tide goes in and out twice a day. In reality, tides go up and down, the in and out effect being produced because the majority of shorelines, at least the ones we visit, are sloping. These sloping shorelines, and the shape of the coastline especially, lead to all sorts of peculiar effects that lessen or magnify the actual rise and fall of the water. Perhaps the most awesome of these are the great bores that rush up certain rivers: the most famous is at Moncton in eastern Canada where the largest tides in the world, with a maximum total rise of 15 metres, rush in over the gently sloping mudflats of the Bay of Fundy and are bottled up as a wall of water that rushes upriver past the town of Moncton.

There is another factor that has a marked influence on the height of the tidal wave crest and that is big brother sun. When the sun and the moon are both in line, their combined gravitational pull acts on the wave crest and we experience the maximum rise and fall of the water, called spring tides. In the same way, when the sun and the moon are at right angles their gravitational effects in part cancel each other out, thus causing a minimum rise and fall of the water, called neap tides. Spring and neap tides alternate throughout the year, each occurring at fortnightly intervals.

The border between the land and the sea thus represents a very special environment that is subject to a series of complex yet ordered changes. Any organism living below low water mark will never have to face the problems of water stress; it will enjoy a temperature that changes only very slowly with the seasons and will throughout its life be bathed in a solution of salts whose composition remains almost constant. Above high water mark things are very different. The environment is anything but stable, being subject to short term fluctuations especially in temperature and the availability of water. Even in temperate climates the surface temperature of exposed rocks can soar up towards 70 °C on a hot, still day, yet in winter when the wind is blowing the effective temperature of the same spot may well be below freezing. Likewise the water supply may vary between supersaturation at the height of a rainstorm to zero during a summer drought or the depths of winter when all surface water is frozen solid. It is therefore little wonder that the denizens of the sea are very different from the denizens of the dry land.

The zone between the tides is a mixture of the best and the worst of both worlds, and it might therefore be expected that the organisms living there must be 'switched on' both to the

A tidal bore rushing up the River Severn in England.

Bladderwrack, a slippery and buoyant brown seaweed.

problems and the opportunities. Apart from the difficulty of hanging on in the teeth of a storm the main trouble with life in this zone, or to use the correct term, the littoral, lies in the fact that although the tidal rhythm at any one spot can be relied upon, the clouds, the wind, and the rain run to no such regular timetable. So just like the organisms living above high water mark, those of the littoral zone must be able to tolerate very rapid changes in their environment. On a hot but stormy summer's day, a plant of the mid-littoral may well be caked with a solid crust of salt, yet a few minutes later be washed in almost pure water. The other unique feature about the sea shore is that here the sea bed is included within the euphotic column but the intensity and quality of the light reaching the bottom will obviously vary with the state of the tide.

Therefore the inshore fringe presents an extensive but harsh environment, open for exploitation by any plant able to overcome the multitude of problems of living there. Many groups of unicellular plants, and especially the diatoms, responded to the challenge and produced forms that became and remained attached to the substratum. However in time three great groups of plants were to become masters of this betwixt-and-between world: these were the green, brown, and red algae, the seaweeds.

The great advantage of becoming a benthic organism (one that lives attached to the ocean bottom) is that there is no longer any problem of staying at the top of the water body. The majority of the plankton expend energy in keeping up near the surface of the euphotic column and the larger and heavier they are the more energy must be used in the 'struggle' to the top. Once firmly attached to the bottom, the prospects for further development are enormous, size and weight being no hindrance. Thus although there are unicellular members of each of the three great groups of the seaweeds, many of them are bulky, being made of many thousands if not millions of cells.

If a plant consists of a single cell, then that cell has to do everything. With the evolution of the eukaryotic or true cell, certain parts of each cell took on specific functions, but the cell remained the complete package deal of life. Once multicellular plants had evolved, there was the possibility of the specialisation of whole cells, in other words, certain cells taking on specific functions in the life of the plant. In the seas of the world there are some 4000 different sorts of large seaweed and the degrees and levels of specialisation are enormous. Among the most complex are the kelps, tangles, and oarweeds, all members of a group found in abundance in both the northern and southern hemispheres, but almost completely absent from the warm waters of the tropics.

Along the Atlantic coast of Europe, *Laminaria hyperborea,* the rough-stalked kelp, forms extensive underwater forests between low water mark and, under ideal conditions, down to about 30 metres. Each plant consists of a very efficient holdfast that, as its name suggests, holds

Kelp forest at low tide, northern hemisphere.

opposite
Kelp forest at low tide, southern hemisphere.

the great plants fast to the bare rock on which they grow. The holdfasts and the thick stipe are perennial and can continue to grow for about fifteen years, by which time the seaweed may be over three metres high from the rock to the tip of the blade. The blade grows from its base and a new one is produced each year, the previous year's gradually eroding away.

A fully grown kelp plant is made up of many millions of cells. All the ones on the outside contain chlorophyll and also other pigments called carotenoids—the pigments that make carrots carrot-coloured and brown seaweeds brown. The cells covering the surface of the kelp not only hold it together, helping to maintain its distinctive shape, but they also carry out the process of photosynthesis and take up the water and salts needed for the life of the plant. In addition, they help to produce the complex of substances that makes all large sea-weeds feel slippery. This 'slippum' is a complex sugar-like substance called alginate.

In order to synthesise it, the cells have to expend energy so why, you may ask, is the precious stuff allowed to leak out into the sea? Living as they do near the shore, the seaweeds would be very prone to damage when lashed about on the rocks, especially at low tide, if it were not for their 'built out' lubricant. Also, during very low tides when the fronds protrude above the water, the alginate rapidly dries to produce a sticky layer that helps protect the seaweed from drying out. The internal structure of the kelp plant is quite complex in that it consists of elongate cells that run down the length of the stalk. In all probability their func-tion is the rapid movement of sugars around the plant, bringing them up from reserves in the perennial parts to speed the growth of the new fronds at the onset of spring.

At certain times of the year brown patches appear on the fronds, and at first sight you might be tempted to think that the frond was begin-ning to die and decay away. Only under the microscope is the true nature of these patches revealed. They consist of tens of thousands of specialised cells, each of which contains 64 tiny cells that are released into the water where they

actively swim about by means of two flagella. These swimming cells are called zoospores and they eventually settle down on solid rock where they begin to grow into a new kelp plant. The intriguing thing is that the new plant looks nothing like the one that produced the spores. It is a tiny filament of cells just a few millimetres long, and it bears in abundance either male or female organs of reproduction. Each of the male organs produces a unicellular reproductive cell called an antherozoid. Pro-pelled by two flagella, this swims through the water to the female organs of reproduction, each of which produces a single non-mobile egg. Fertilisation is accomplished by fusion of the two reproductive cells and the resultant zygote divides again and again, growing to produce new baby kelp plants that are miniature replicas of their original parent.

One adult kelp plant can produce tens of millions of zoospores, each of which is capable of growing into a new reproductive plant. In turn, each tiny plant can produce many hun-dreds of eggs, and if they were all to come to fruition the sea would soon be solid with rough-stalked kelp plants.

Of course this never happens because there are problems for the kelp all along the way. The zoospores may be eaten before they settle down to grow, many of them may be swept into deep water where the bottom of the sea receives insufficient light for the growth of photosyn-thetic plants, or the tiny reproductive plants may be grazed off by a horde of ravenous animal life before they reach maturity. Even the new generation of kelp plants is not out of trouble because it is very likely that it will be grow-ing beneath a thick canopy of adult kelps where again there is insufficient light to allow develop-ment. The sporelings can 'hold on' in a sus-pended state of growth, but only for a short period of time, so unless an adult canopy plant dies and allows light into the depths of the forest, the sporeling will not survive. Thus as soon as a gap opens up in the forest canopy it is filled by the rapid growth of a sporeling that has been waiting its chance down in the gloom. In this way the world population of kelp plants is regulated and the structure of these underwater forests maintained.

One of the great problems of becoming fixed to the bottom is that of dispersal, the age old problem of the progency escaping from the competitive influence of their parents. For this reason the majority of benthic organisms, including the kelp, spend at least part of their life cycle up among the plankton, and so gain the advantages of dispersion by riding the currents of the oceans.

Along the Pacific coast of California the largest kelp of all, *Macrocystis pyrifera,* grows in abundance. It is an annual and can grow a staggering 50 metres in length in one season, the fronds being buoyed up by small air sacs at the base of each leaf-like branch. The giant kelp is a good and very prolific producer of alginates, and because alginates are important raw materials to such a wide range of things as cosmetics, food, and chemicals, the giant kelp forms the slippery basis of a multi-million dollar industry.

In the early 1950s all was not well along the coast of California: some of the most productive beds of giant kelp gradually began to go out of production, and the plant began to disappear. Was it overexploitation? Or was it the increasing pollution that cut down the amount of light penetrating to the all-important reproductive and sporeling stages? A team of scientists headed by Dr Wheeler J. North dived into action to try to solve the riddle of the disappearing kelp. Their study singled out one main culprit, a spiny sea urchin that feeds voraciously on the plant, chewing through the multiple stalks just above the holdfast. The question then became: why do they only destroy the kelp forests in polluted waters? The answer was a complex one. Coastline pollution was benefiting the sea urchin population by providing an alternative source of food (they are omnivores, that is, they can eat just about anything that comes their way). At the same time heavy overfishing of the waters was effectively removing the fish that themselves feed on the urchins. The case of the disappearing kelp thus proves beyond doubt that you cannot tamper with any one

part of an ecosystem without affecting the delicate balance of the whole living complex.

The brown seaweeds may boast among their number the largest of all marine plants but the prize for beauty must go to the red seaweeds, with some of the more delicate greens running a close second. All three types contain chlorophyll, for without this vital green pigment photosynthesis is impossible. In addition, the brown ones contain carotenoids, and the reds a complex of red pigments. These other pigments are called accessory because, although they cannot replace chlorophyll as the cornerstone of photosynthesis, they can and do absorb other wavelengths of light and pass them into the process via chlorophyll.

Take a close look at a large pool situated at about mid-tide mark on a rocky shore. All three types of seaweeds will be there in all their colourful glory and usually they will be arranged in a more or less ordered pattern. The green seaweeds will be most abundant around the rim of the pool; further in the large browns will predominate; while the red seaweeds will increase in abundance with depth. A similar zonation is often found passing from the upper tidal limit down to and below low water mark. The green seaweeds (see page 35) predominate first, with thick beds of browns down to low tide mark. Just below the permanent water mark the kelp forests still predominate, but with increasing depth the delicate red seaweeds (see page 36) of the forest floor become more abundant. Whether the zonation is related to the fact that the greens are less susceptible to the effects of dessication than the reds, or because the red pigment gives the latter an advantage as it can absorb the blue wavelengths of light that predominate in the deeper water, is still a matter of debate and experiment. The answer is probably a combination of both, plus perhaps a few other factors.

The study of the 'life style' of a single plant is called autecology, and the autecology of seaweeds is complicated by the fact that there are at least two parts of the life cycle to be taken into consideration. Often one phase of the life cycle is a large, easily recognisable seaweed, whereas the other phase consists of no more than a few cells growing over the rock surface. One of the tasks that faces any would-be seaweed autecologist is to unravel the life history of the chosen plant, and some of them can be very complex. In recent years many of the world's experts at unravelling the life cycles of seaweeds have been women; a few names are Mary Park, Lilly Newton, Elsie Burrows, and Connie MacFarland. Perhaps it is the patience needed for this type of work that makes it the domain of the lady scientists? Who knows, but painstaking work it is. If you are saying to yourself (or just thinking) 'why bother, what a waste of time', then read the following story.

In Japan, one seaweed called nori, or *Porphyra*, is highly cherished as a food plant. It has been cultivated for a long time in massive farms, which today cover thousands of acres and have an annual output of more than 150,000 tonnes. The seaweed is grown on nets suspended at just

Turtle grass meadow in the Mediterranean.

Nori or laver bread, an edible red seaweed.

the right level in the water. In the unenlightened days, before the intricacies of the life cycle of *Porphyra* were known, the main hang-up in nori culture was the seeding of the nets. The discovery by Kathleen Drew that a microscopic red seaweed, which lives by boring into mussel shells, was the other half of the life cycle of *Porphyra*, helped to solve the problem. So grateful was the nori industry that in her memory it erected a monument that overlooks the 500 kilometres of nori nets strung each year along the edge of Tokyo Bay.

Apart from the zonation of the main types of seaweed across the tidal range at any one site, they are also zoned on a world scale. The large browns are very abundant in the temperate and subarctic latitudes. In contrast the reds reach their peak of importance in the tropics, whereas the greens appear to enjoy the water whatever the temperature. The exact reason why the large browns have missed out when it comes to

Close-up of eel grass.

life in warm tropical waters, we do not know, but the fact is that in the tropics they are only conspicuous by their absence.

The evolution of the large seaweeds has also missed out in another way. Their holdfasts are absolutely perfect for solid rock but they are of no use at all when it comes to life on the shifting sands and silts that typify many of the coastlines of the world. For this reason great stretches of the world's inshore zone must have been devoid of large plant life for a very long period of evolutionary time. But now the situation is quite different. There are a number of large plants collectively known as the turtle grasses that grow luxuriantly, covering the sands and silts of the littoral and sub-littoral zone.

Turtle grasses are neither seaweeds nor, despite their name, are they grasses. Their leaves, like those of all true land plants, have veins, and at the base of the mass of leaves is not a holdfast but an intricate system of roots penetrating down into the silt on which they grow. A close look at the right time of year will also show that the turtle grasses bear underwater flowers which, just like their dry land counterparts, produce pollen grains and set seed.

The turtle grasses are closely related to the pond weeds that inhabit bodies of fresh water. It would thus appear that the potential of these underwater meadows remained untapped until evolution had produced a group of plants fitted to the rigours of life on land, one of their key adaptations being a root system able to penetrate down into the dry soil. Subsequently some of them must have re-adapted, overcoming the problems of life underwater, and moved back into the sea to make use of those wide open, sandy places. Today the turtle grasses are very widespread. They are found in all the large oceans, but are especially abundant in the warm waters of the Mediterranean, and are important members of the complex communities of tropical coral reefs.

So it was that in time the plant kingdom evolved to make full use of the shallow parts of the sea—a mantle of green, brown, and red transforming part of the incident energy of the sun into the complex of organic chemicals called plant life.

forms
of life

The four grades of construction

As the plant kingdom evolved to fill the many niches in the sea, fixing more and more of the energy incident on the surface of the planet, the multitude of evolved forms presented new potential for evolution. This was potential in the form of energy-rich chemicals that could in turn be used by other organisms, the organisms we now call animals. Like the plants, the animals had to overcome problems, both related to the environment and to life within the evolving milieu, that is, to the problems of living with each other.

Modern-day science records approximately 350,000 different types of plants and well over one million different sorts of animals. This fact would not only present a problem to a present-day Noah whose ark would have to be capable of holding a viable population of each, it creates a real and just as vital problem to the students of the biological sciences.

It would be absolutely impossible for any one person to accomplish the task of learning to recognise at sight either all the animal or all the plant species present in the world today. To get to know a single group in depth is often the work of a devoted lifetime. One great student of the animal kingdom, Dr Libbie Henrietta Hyman of the American Museum of Natural History in New York, set herself the prodigious task of not only obtaining a working knowledge of the animal kingdom, but of writing a text in which she aimed to cover the most diverse of the groups, the animals that have no backbones.

The work unfortunately was never completed, but the six volumes we do have rank among the most detailed accounts of the products of evolution to date. However they are books for the specialist, and anyone without a detailed training in the subject would soon be lost in the plethora of accurate fact. But it must be stressed that these are not static, descriptive works. They are alive with the fire of evolution and are a flowing account of how life overcame every obstacle to reap the full advantage of each environment. In these great works Libbie Hyman captures the essence of evolution in the concept of the four 'grades of construction', each grade carrying the living state further from the simplicity of the unicell through successive stages of complexity: acellular, cellular, tissue, organ.

An acellular living body is one undivided by cell walls, the single cell being the total functional unit of life. Even when such cells are aggregated into colonies they all have the same form and total life function throughout most of their lifespan. All the protists are included in this the simplest grade of construction. The cellular grade is reserved for the sponges, which consist of large numbers of cells of a number of different forms and functions, loosely aggregated into a recognisable whole. Animals with tissues make up the next grade of evolutionary endeavour. Tissues are aggregates of similar cells all performing the same function or functions in the life of the organism. This group includes the jellyfish, sea anemones, corals, and the comb jellies. The rest of the animal kingdom including man comes into the organ grade category. Organs are functional units made up of a number of tissues organised to perform one particular function.

With each upgrading the animals became more complex and a new range of habitats were

Soft coral *Goniapora* with polyps fully extended.

opened up to the animal kingdom. To house and protect the living chemicals, evolution was producing a structured, internally controlled environment within which the chemistry of life was further and further divorced from the direct effects of the fluctuations of the environment outside.

With the increasing complexity of the eminently portable module for life, certain problems arose relating both to life and to reproduction. The unicell was engendered with the immort-ality of total regeneration, but as complexity increased, the practicality of regeneration became less, and the function of reproduction was restricted to certain cells in specialised organs. The functional units of life—the animals we see—are biologically speaking nothing more than a vehicle for producing the germ cells of the next generation: 'a chicken is just an egg's way of making another egg'.

59

A 'whole' among friends
sponges

Sitting in a lovely hot bath with a luxuriantly expensive sponge, it is hard to come to terms with the fact that you are squeezing the remains of a living organism. The problem today is that more often than not your companion of the bath is no more than a plastic imitation of the skeleton of this remarkable animal. Real, large bath sponges are rare and costly things, and they are getting rarer mainly due to excessive exploitation—could the right word for this be 'oversponging'?

Sponges are remarkable animals in many ways but their most remarkable feature is their ability to regenerate. Take a sponge, smash it up, and squeeze the remains through fine silk,

Sponge opened up for inspection.

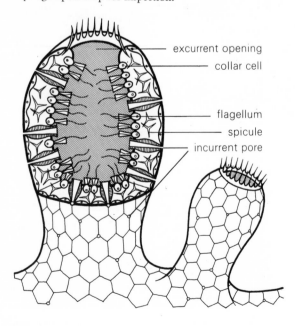

excurrent opening
collar cell

flagellum
spicule
incurrent pore

and the result will be an extrusion of cells capable of regenerating by forming clumps and unions that together grow into a whole army of new sponges. However the process takes a long time and requires good clean water conditions and a ready supply of food.

Adult sponges are no more than a loose aggregation of cells held together, or rather apart, by holes. The bulk of the cells that make up the body are very similar, being rather like amoebae, and they can move about performing various functions around the structure of the sponge. On the outside they take on a flattened shape that adequately covers the surface, a surface perforated by many thousands of tiny pores. Each pore is surrounded by a cell that looks not unlike an elongated American doughnut having taken on the exact shape of the pore. What is more they are able by expansion and contraction to control the size of the pore and hence the ingress and egress of water.

Rightly enough, the collective name of the sponges is the *Porifera*, meaning pore bearers. The most specialised of the component cells are the collar cells, which line the internal cavities. Each one has a collar-like expansion encircling the base of a single long flagellum that extends into the internal cavity of the sponge. It is the beating of these flagella that draws water into the pores and through the body of the sponge, but the remarkable thing is that they do not

opposite
A cup sponge.

overleaf
A large sponge, the home of the blue-head fish.

60

beat in phase. Sponges have no nervous system and so the activity of the separate cells is by and large completely uncoordinated. It is thus still something of a mystery as to how the random waggings of many millions of flagella can maintain a constant flow of water in through the tiny inlet pores and out through the larger outlet pores that typify this weird group of animals. Yet the fact remains that a large sponge may shift many hundreds of litres of water in one day. This current of water supplies the sponge with both the oxygen and the food it needs and carries away any waste products out into the surrounding sea.

Not only do the collar cells maintain the all-important water currents but they also trap the food particles, engulfing them in much the same way as a free-living amoeba. The pathways and mechanisms of food uptake have been studied by feeding harmless particulate dyes into the water stream and tracing their passage through the cells of the sponge. As soon as the food particles are inside the cell the process of diges-

tion begins. The partly digested food may however be passed on to other less specialised cells that are totally amoeboid and can move throughout the sponge, passing the end products of digestion, energy-rich compounds, and waste to wherever they are required or can be dealt with. Again such movements must be somewhat haphazard as there is no system of coordination, either internal or external. The wandering cells can also settle down and take on the more specialised function of secreting the spicules or fibres that go to make up the sponge's skeleton.

These spicules may be made of chalk or silica, both of which are borrowed from the sea water and laid down by the cell, or of spongin, a structural protein that is manufactured by the cells themselves. The type of skeleton forms the basis for at least part of the classification of the sponges, which are grouped into the calcareous, the glass or silica, and the horny.

The calcareous sponges are both common and abundant in inshore waters where they appear as a somewhat bristly, drab coloured

overleaf
A plumrose anemone open to callers (see pages 79–81).

opposite
Coral skeleton with retracted polyps. The green is chlorophyll (see page 85).

A sponge spicule (magnification × 3750).

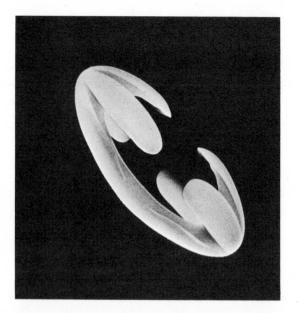

Assorted sponge spicules (magnifications: above and
right, × 1800; below, × 2400).

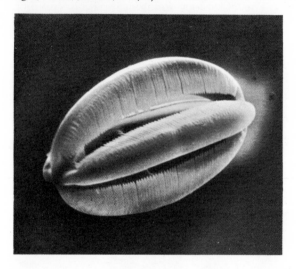

sponge plumbing indicate that it is the structure
and arrangement of the batteries of internal
chambers that help to turn the random movements of the flagella into a continuous one-way
stream of water. Structure helps to create order
from disorder, and part of this structure is due to
the manner in which the various spicules are
laid down in formal array. Again, just how the
disorganised secretory cells attain this degree of
organisation of their products is not known.

Of all the groups it is the glass sponges that
show the highest degree of order when it comes
to the arrangement of the spicules. They are all
very special, each being based on the complex
format of having three axes crossing at right
angles. These complex spicules are called
hexactines, and the scientific name for the glass
sponges is the *Hexactinellida*. The ordered
siliceous skeleton supports a complex of chambers each lined by collar cells. Their skeletons
are avidly collected as ornaments for they must
be classed among the most intricately beautiful
of all natural products.

In the course of the life of a sponge many
living things get drawn in with the water stream
and some, finding an equable environment well
supplied with oxygenated, food-rich water,
take up residence. The only problem is that it
can become permanent if they grow too big to

of the most complicated forms: these consist of
complexes of branching tubes that anastomose
and fuse together into a great mass of chambers,
each of which is lined by batteries of the flagellated collar cells that form the core of every
sponge. Recent studies of the hydrodynamics of
covering on large underwater objects. They
include among their number some of the
simplest sponges, consisting of no more than
an aggregation of bags each with a single large
exit pore at the top. But they also include some

escape through the outlet holes. Glass sponges can thus become living cages for a whole variety of species, especially shrimps and crabs, and these then have no choice but to live out their days enmeshed in finest silica. The skeleton of the Venus' flower basket sponge with an imprisoned pair of shrimps is highly prized in Japan as a wedding present for it symbolises that marriage is unto death.

In many cases the inhabitants of the sponges are not there just by chance. With time, special relationships have developed between the sponge and its permanent residents. Certain shrimps are found living almost exclusively in particular types, and a large sponge may contain many thousands of individuals of one species of squat-

Another sponge spicule (magnification × 9600).

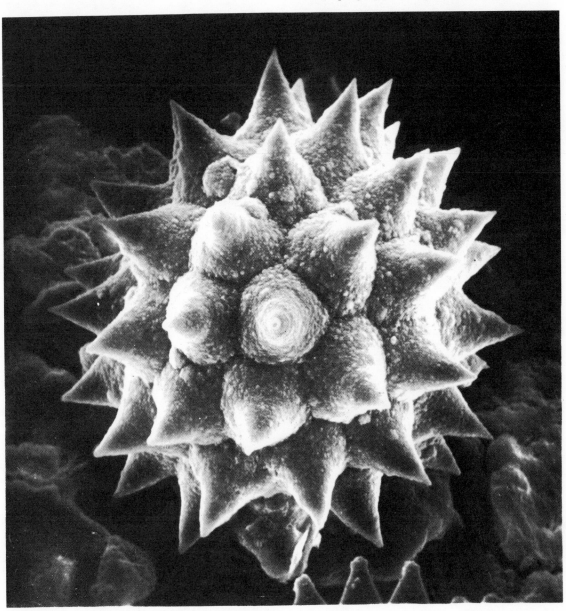

ter. This kind of relationship, where one animal makes its home inside another, is called commensalism.

An extreme form of this home-making is found in a group of sponges that appear to grow only on the outside of old mollusc shells that are themselves already the home of a hermit crab. In time the sponge completely covers both the outside and the inside of the shell, which eventually disintegrates and disappears. However the hermit crab is apparently unperturbed and it continues to live within the protection of its new walls.

Bath sponges are members of the third group, those whose skeletons are made of spongin. It is this proteinaceous matrix that, after drying in the sun, is kneaded and repeatedly washed before you use it in your bath, along of course with the most expensive of toilet soaps. For obvious reasons neither of the other groups can do the job of the bath sponge, although they might make good stand-ins for scrubbing brushes or loofahs. Loofahs are not sponges at all but the dried skeleton of the fruit of a plant that looks like (indeed, it is not unlike) a cucumber.

The wandering amoeboid cells that make up the bulk of the living body of all three types of sponge have one other function, and that is sexual reproduction. Despite their amazing powers of direct regeneration, the sponges do produce from these cells typical male and female reproductive cells—mobile spermatozoa and non-mobile eggs. The spermatozoa are re-

opposite
A Venus flower basket.

The purse sponge *Grantia compressa*.

leased into the water stream that flows through the parent sponge to be carried into other nearby sponges where, if ripe eggs are present, fertilisation may be effected. Many variations on this theme have been described but the end result is a fertilised egg that, while still within the parent, develops into a ball of flagellated cells called a larva. The larva then works its way out into the water current and hence into the open ocean where, after swimming about, it develops into a perfect miniature of its parent.

Although the bath sponge is now becoming a rarity both in and out of captivity, the *Porifera* are very common denizens of the sea. They are

especially abundant in shallow inshore waters where they often grow as brightly coloured crusts covering larger objects. They have few natural enemies, possibly because apart from their surfeit of spicules they all appear to have a disagreeable taste and smell, and some have been shown to be poisonous.

If you were to look for a good candidate to link the protists with the more complex multi-celled animals, the sponges might appear to be an obvious choice: they are aggregations of single cells that, although lacking any control system, act together as an at least partially integrated whole (pardon the pun!). Perhaps

Halichondria panicea, the breadcrumb sponge.

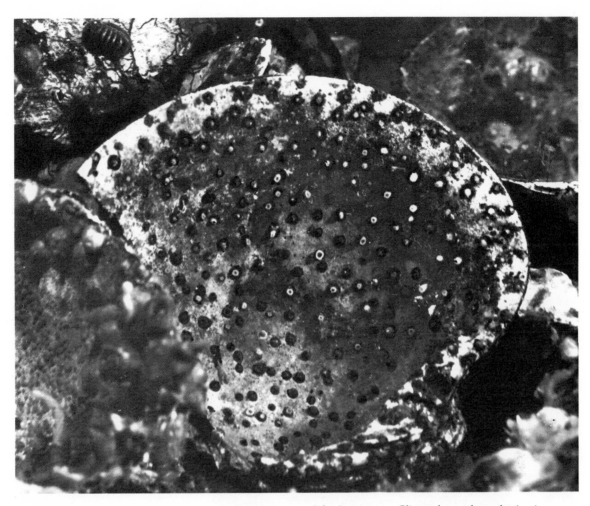

A boring sponge, *Cliona celata*, at home boring into a cyprina shell.

the best description would be that a sponge is a colony of cells, certain members of which are specialised for certain jobs, yet each one is capable of going its own way in life.

Libbie Hyman regarded them as being something more than a simple colony and therefore places them in the cellular grade of construction, for they are more complex than the protists but less organised and specialised than the next grade (in which the cells are aggregated together into definite tissues whose workings are coordinated by means of special cells). The *Porifera* probably do not represent a stage in the main stream of evolution but rather a side shoot that has exhibited its disorganised success in the oceans of the world for at least 1000 million years.

Variations on a theme
jellyfish, sea anemones, corals

I often wonder what it would be like if the human eye had the power of resolution of a good light microscope. A swim in the sea would certainly be a novel experience, gingerly elbowing your way through a host of mini-monsters each going about his business. The plant plankton would at least be coloured and therefore much easier to see than their animal counterparts, which in the main are transparent but for their eye-spots and the plant food they are busy digesting. You would be surrounded by ghost-like shapes set against a backdrop of fine green mist, a mist composed of the smallest of the plant plankton, the microflagellates. Even the new 'super eye' would have difficulty in resolving these for, when fully grown, they are only about 0.005 millimetres in diameter—

a myriad of minute cells swimming actively through the water, lashed along by their flagella.

It would not only be a world of moving beauty but also a world full of the drama and the struggle for life, where survival for one usually means death for many others. Perhaps the nearest one can come to this idyllic (or perhaps nightmarish) situation is to dive among a shoal of jellyfish. The beauty of their transparent symmetry against the interplay of light is almost beyond description—as is the agony if you are unlucky enough to come across one of the many stinging varieties.

Jellyfish are members of a world-wide group of animals called the *Cnidaria*, a name meaning 'the ones with the threads', and it is the threads

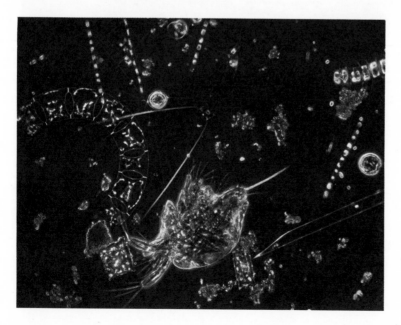

Assorted plankton, minimonsters making a world of beauty (magnification × 1800).

opposite

A gas bag with a trawl of death: the dreaded Portuguese man-of-war.

that do the stinging. They are megaplankton and include in their ranks some of the largest animals that float free in the currents of the oceans. The largest, *Cyanea arctica*, is a wonderful shade of blue and can grow to more than two metres across its umbrella.

All true jellyfish consist of a bell, or umbrella, that is usually transparent and always has a basic tetramerous symmetry (divisible into four similar quarters). The structure that spoils the true radial symmetry of the bell is the four lobes surrounding the mouth. The mouth leads to a large cavity, the 'gut', which is lined with a layer of cells whose special function is to help digest the food. The outside of the umbrella is similarly covered with a layer of cells specialised to protect and in part support the bell. Between these two definite cell layers or tissues is a layer of non-living jelly making up the bulk of the animal. Although the bell looks not unlike a semi-rigid polythene bag, it is in fact a remarkably solid structure. This can easily be verified by testing with your foot one that has been stranded high and dry on the beach; but a word of warning: keep your shoes on because it could be one of the stinging variety. Better still, lift it up very carefully with your bucket and spade and put it back in the sea where it belongs.

The stinging cells are the unique feature of the cnidarians. They may be located on either side of the bell but are especially abundant on the lobes around the mouth and on the tentacles that fringe or hang down from the umbrella. The tentacles give the bell its proper name of medusa, named after the mythological creature whose head was crowned with venomous snakes. Like the animals that bear them, the stinging cells come in all shapes and sizes. Each consists of a single flask-shaped cell inside which is a hollow thread made of protein, and each is equipped with a short trigger hair protruding from its top. Anything bumping against the trigger can fire the cell: the coiled thread shoots out, turning inside out as it goes, but its base remains attached to the cell. Seventeen different types of stinging cell have been des-

cribed and all have been given complicated names, one of the best being 'heterotrichous microbasic eurytele'! Whatever they are called they are all highly efficient. Some of them coil

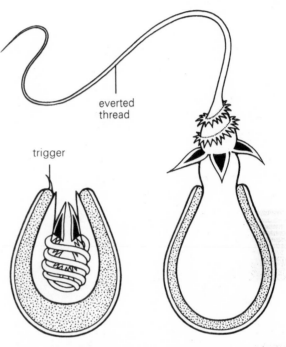

Stinging cell before and after blast off (above), and a battery of stinging cells, all non-returnable (below): A = Telotrichous macrobasic eurytele; B = Heterotrichous microbasic eurytele; C = desmoneme (undischarged); D = stenole; E = Microbasic mastigophore.

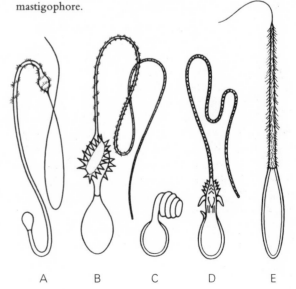

A   B   C   D   E

opposite

Agalma, a complex siphonophore.

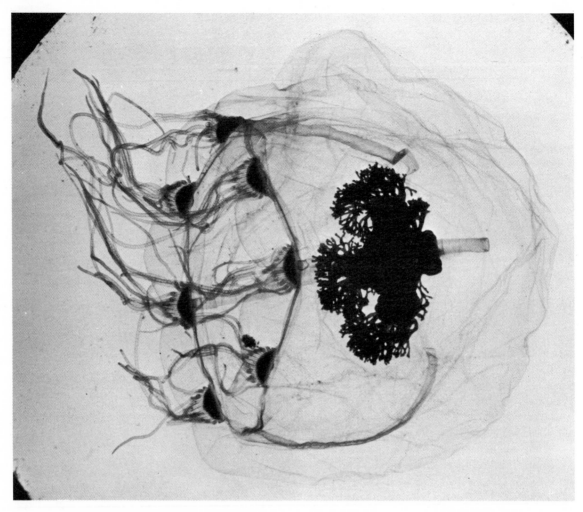

A tiny medusa jellyfish *Lizzia koellikeri* (magnification × 12).

around the prey, others stick to it, while the most deadly penetrate its body, injecting poison through the tubular thread.

Apart from their great complexity there are a number of rather remarkable features about the stinging cells. First, they are non-returnable: once fired they cannot be used again and are therefore cast off. One meal could mean the loss of many thousands of stinging cells, representing a considerable drain on the jellyfish's energy. Therefore some of the energy obtained from the meal must be immediately channelled into the manufacture of new cells—a very inefficient process. Second, the new stinging cells are not formed *in situ*. They begin their

development at a point far removed from their final strategic position and then slowly migrate, while still growing, through the body of the animal to take their appointed position on the firing line. Third, experiment has proved that the trigger hairs are in the main activated mechanically and yet when the jellyfish is fully fed it is very difficult to make the stinging cells fire their threads. Just how the animal achieves this control is not known.

Despite their small size, not all the stinging cells are harmless to man. Reaction to their poison ranges from nothing, through a slight burning sensation, to severe effects leading to convulsions and, in some cases, death. The worst offenders are members of a group called the cubomedusans. As their name suggests they are the squares of the medusoid world, their swimming bells having a box-like outline. Squares they may be, but some of them pack a deadly punch and so are a great hazard to swimmers, especially in the warm waters of northern Australia where they abound.

Jellyfish swim by means of rhythmical pulsations of their umbrellas which are well supplied with contractile muscles formed from the outer layer of cells. In this way they manage to keep their transparent bulks up in their chosen habitat. For the majority, this is close to the surface of the sea, although a few do prefer deeper water. All of them have mastered the art of trawling—

they float motionless, spreading out a great net of instant poison. Once paralysed by a broadside from thousands of stinging cells, the now quiescent prey is drawn in towards the four-lobed mouth, which is often hidden among great flocculent masses of coloured tissue that is itself well armoured with yet more stinging cells.

Their food partly depends on their size, but small fish are certainly a favourite of medium to large specimens. Certain fish have developed a special relationship with certain medusae and find, if not their homes, their place of retreat among the poisonous tentacles. In many cases it appears that these fish are not immune to the poison but have the ability of not triggering

Lucernariopsis campanulata, a stalked jellyfish.

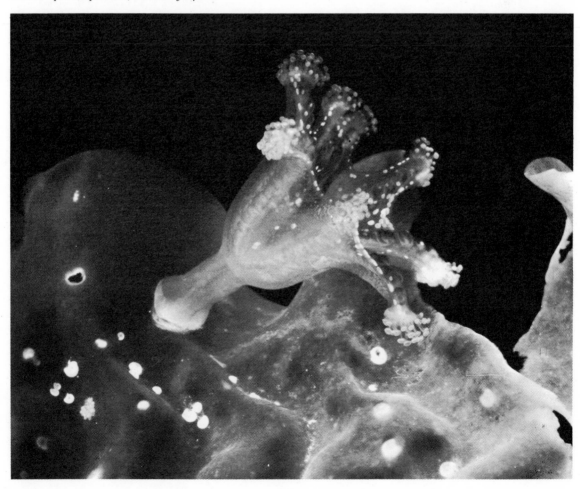

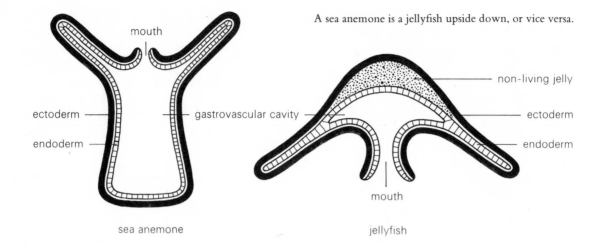

mouth

ectoderm

endoderm

gastrovascular cavity

sea anemone

A sea anemone is a jellyfish upside down, or vice versa.

non-living jelly

ectoderm

endoderm

mouth

jellyfish

off the threads. It is easy to see what the fish get out of the 'friendship' but difficult to understand how the jellyfish benefits, especially now it is known that some of the ungrateful fish are not averse to feeding on the tissues of their host. The commensal fish may of course act as decoys, luring other fish within trawling range. This unlikely pair make a fascinating sight as they swim along together, the fish ready to duck into home as soon as danger approaches.

First prize to *Tealia felina,* the dahlia anemone.

Without doubt jellyfish are animals. They are made of many cells, aggregated into two layers, that are each specialised to do a particular job. The rhythmical swimming motions and combined attack of all the tentacles when feeding show that the activity of the cells and tissues is coordinated. This coordination is brought about by special nerve cells that form a loose network throughout the tissues.

Their bodies also contain many complex aggregations: each comprises several different types of cell, of which some are sensitive to light, others to gravity. The light sensors consist of a lens that focuses the light onto pigmented receptor cells. The gravity sensors are even more complex. They have a central body that contains a large proportion of chalk. This is suspended in a cavity through which it falls under the force of gravity to lie on the receptor cells when the animal is the correct way up. Both kinds of complex sensors are well supplied with nerve cells that feed impulses to the nerve net, and in this way the movements of the whole animal may be correlated to keep the trawler upright and on station over its fishing grounds.

Jellyfish belong to the third or tissue grade of construction. But their possession of definite sense organs, that is, groups of different tissues that work together to perform a specific function of importance to the life of the whole animal, points to the next, more complex grade. On the other hand the jellyfish and their rela-

Metridium senile, a plumrose anemone (see also overleaf and page 63).

tives show some links with the past, especially in the fact that those all-important stinging cells spend at least some of their short life wandering through the body of the animal. The cnidarians would thus appear to be a halfway evolutionary house between the cellular and the organ grades of construction. They must not be regarded as a 'missing link' between the sponges and the more complex forms, but simply as an experiment in evolution from which we may draw inferences as to the mechanisms of the overall process. And it was a successful experiment at that, for their members abound in every ocean and in all the major ocean environments.

The transparent jellyfish, pulsating its un-hurried way through the surface currents, seems a far cry from the stony corals that build

the great reefs sheltering tropical shores, yet they are very closely related. Furthermore, turn a jellyfish upside down and fix it firmly by the top of its bell so that its tentacles wave in the air, and you have almost got a transparent sea anemone. Sea anemones, jellyfish, and corals are all linked by the common threads of their stinging cells.

Turning upside down is precisely what one

A plumrose anemone shutting up shop (above and opposite).

enterprising medusa does do: at sea, a young *Cassiopea* is just an ordinary jellyfish trawling its life away. When it gets into shallow, sheltered water, instead of being washed up and stranded ashore, it simply turns over and becomes fixed to the bottom by means of a raised ridge of

tissue that encircles the edge of the bell. In this position it can feed effectively, the pulsations of its bell 'pumping' in a continuous supply of food. These upside-down jellyfish can aslo feed in another and rather remarkable way. Inside their translucent tissues are many unicellular plants. *Cassiopea* provides a home for the plants which in turn provide the host with oxygen and probably with sugars, both the

products of photosynthesis. Such a relationship of mutual aid is termed symbiosis, and such symbioses are of common occurrence in jellyfish, anemones, and corals.

A true sea anemone consists of a stalk-like body whose bottom end is modified for attachment. At the top a ring (or rings) of tentacles surrounds the central opening that leads into a blind sack. The opening, like that of the jellyfish, must serve both as a mouth for the intake of food and as an anus for the egress of any indigestible parts of the prey. Digestion takes place in two stages: enzymes secreted from the gut begin the process, and the partially-digested food is then taken up into the cells of the gut wall where the process is completed.

For a long time the sea anemones, as their common name suggests, were reckoned to be plants and some bear the name of the flowers they so much resemble, for example, the dahlia anemone. Their tentacles may be petal-like but each is armed with efficient stinging hairs, the trade mark of the group. In the main they are solitary, each polyp going its own free way. Although their movements are very slow they can and do move about by gliding along on their adhesive foot. Even the large ones (and some grow into real monsters, more than one metre tall) can move ponderously across the sea bed.

Anemones have been the subject of many experiments aimed at obtaining an understanding of the workings of their simple nerve net. Careful stimulation of a single tentacle at one side of a complex crown brings about an ordered response of the adjacent tentacles, with the area affected depending on the intensity of

The snakelocks anemone which cannot withdraw its tentacles.

opposite
Shy damsel fish safe at home among the stinging cells.

the stimulus. Food causes an immediate reaction: the tentacles move towards the prey and fire off batteries of threads on contact. Damage, on the other hand, leads to rapid contraction of the tentacular crown.

Corals are anemones that lay down massive chalky skeletons outside their actual tissues. Each coral polyp thus sits in its own private pulpit, into whose confines it can rapidly withdraw. The lamellae radiating out from the margin towards the centre of the pulpit correspond to the spaces between the internal plates of cells that divide up the polyp body, rather like the slices of a cake. Like the anemones, the coral polyps have a hexametric symmetry. There are however usually more than six septa, and the complexities of the internal divisions are of great delight to zoologists. Suffice it to say that the arrangement of the septa is an important feature in the classification of the corals and one that adds much to the beauty of their intricate skeletons. It is these skeletons, together with the chalky remains of many other types of animals and plants, that have in time gradually built up the reefs of the tropical seas (see also pages 258–66).

The corals range from solitary forms like the amazing *Fungia* which despite its massive skeleton can move about on the bottom, to the delicate branching acropore in which each polyp occupies a raised cup on the surface of the main skeleton, and on to the massive brain coral whose sinuous patterns are occupied by series of con-

Soft coral with expanded polyps

tiguous polyps that share a common fringe of
tentacles. In the same way that the tissues of
Cassiopea form the home of many thousands of
unicellular plants, so too do the polyps of the
reef-building corals (page 64). There is much
evidence to show that, without their symbionts,
corals cannot form their skeletons. This accounts
for their restriction to locations above the 50
metre mark and for the fact that the bulk of reef
building takes place in waters shallower than
30 metres.

Not all corals are reef builders, and conversely
a living reef front is a mosaic of many other sessile cnidarians: soft corals, sea pens, dead men's
fingers, sea whips, and gorgonians. All these are
closely related to the corals but they differ
mainly by the fact that their polyps have eight
internal septa and by the lack, or type, of
skeleton. Some of them do produce a massive
skeleton, and the best known is the bright red
organ pipe coral. In the clear waters of the central
Indian Ocean, below the 50 metre mark, the
reef-forming corals give way to a zone dominated by groves of gorgonians and sea whips,
some of which are two metres tall. Typically
the large colonies of the gorgonians, whose
tough, horny skeletons are made of a special
substance called gorgonin, expand their netted
fronds in one plane so that they look not unlike
the skeleton of a gigantic leaf. This is also the
realm of some of the larger anemones, which
glide about in a meadow of glass sponges. The
zone that girdles the base of each reef is one of
the most beautiful, most mysterious, and most
painful habitats to visit on earth. Everything that
does not sting seems to possess either sharp spines
or needle-like projections and here, beyond the
protection of the coral reef, the great predatory
fish always seem to be so much more menacing.
So it is that the cnidarians have the best of both
worlds—the medusae trawling the potential of
the planktonic world and the polyps 'living it
up' on the bottom.

There is another group of cnidarians of which
many lead a dual life: for part of it they exist as
a free-swimming medusa and for part as an
attached polyp. This group is called the *Hydrozoa*
and its members are abundant in most waters

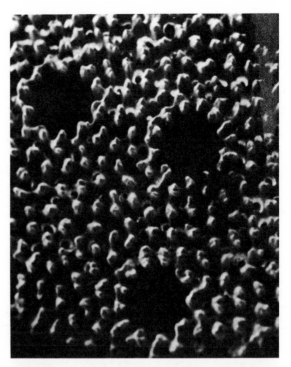

Closer and closer to the limestone pulpit in which the
coral polyp sits (magnification × 40 and × 100).

Clava squamata, a colonial hydroid (magnification ×2.4).

but as many of them are quite small they tend to be overlooked. Again the easiest ones to find live in tropical waters, and although they possess large calcareous skeletons, often coloured blue, they are best located by feel, not sight. These are the 'fire corals' and because of their fiery stings they are the bane of every diver's life. The proper name for these reef builders is the millepores, the word referring to the many tiny pores that pit their chalky skeleton and from which the small polyps project.

Examination reveals three types of polyp. Those that feed the colony have a typical ring of armed tentacles surrounding a gut cavity that is continuous with the rest of the colony. Defensive polyps have well-armed tentacles but no mouth, so they are unable to feed themselves. The third type occupy somewhat larger holes and each one is like a much reduced medusa. These are released into the sea where they live

for only a few hours before dying and disintegrating but during this short free existence they release the male and female sex cells into the sea for fertilisation to take place. A small larva then joins the mini plankton before settling down to grow and bud into a new colonial animal. The fire corals are therefore polymorphic —three different types of polyp each perform specific functions for the colony. They also exhibit a marked alternation of two generations: the first generation, a fixed colony of polyps, buds off the second, which consists of tiny planktonic medusae that bear the sex cells and complete the life cycle. However the most extreme cases are found in the smaller but much more diverse hydroids that form a stiff, branching, felt-like covering on the surface of large seaweeds and rocks.

Hydroids vary enormously: some have a tough outer covering, others are naked, and they range from solitary forms to intricate feather-like colonies. They all have a range of polyp types and some bud off many tiny medusae. Often these are free living, being capable of feeding themselves for a considerable time, and thus their function is both dispersion and reproduction. The product of fertilisation develops into a flagellated larva that swims with the plankton before becoming fixed to its appointed spot.

The weirdest of all the cnidarians is a group closely related to the hydroids and millipores that has forsaken life fixed to the bottom and has carried polymorphism to its extreme up among the megaplankton. The group, which has no common name, is called the siphonophores, although a good name for them might be the bellbuoys. The best known member of the group is undoubtedly the much feared Portuguese man-of-war.

Like all the siphonophores this terror of the seas is a bag, not itself of trouble, but trailing a net that is full of trouble and up to 20 metres in length. At the base of the bag, which is a modified medusa, is a gland that secretes gas into the

opposite
Tubularia indivisa, a hydroid with stinging cells at the ready (magnification ×7).

Other members of the siphonophore group lack the apical float, and instead have modified swimming bells and varying forms of float from which trail a bizarre mixture of feeding, defensive, and reproductive polyps, to the extent that the whole thing looks like some inflatable, transparent, nightmarish Christmas tree. At the onset of stormy weather, some are able to deflate their floats and, with the aid of their highly muscular swimming bells, literally dive out of trouble. But not so the blue or pink Portuguese man-of-war. It is at the mercy of the winds, which accounts for the great flotillas seen out at sea and, more often, found stranded on the beach in their thousands.

As with the colony of cells making up the body of the sponge, the colony of polyps and medusae that constitute a siphonophore work

Plumularia halecioides, an ordered colony of polyps.

Organ pipe coral skeleton showing the successive layers of pipes.

bag thus keeping it inflated. Below this, on the main stalk of the colony, are arranged a truly ferocious army of individuals of varied shape, form, and function. Feeding polyps, with long fishing tentacles that may be quickly 'reeled' in, are situated among many protective polyps, each with more than their fair share of stinging cells. It is the form of the tentacles that sets the siphonophores apart from all the other members of the group, for the armed polyps have single, long, highly contractile tentacles that may themselves have contractile branches each ending in a knob, or spiral, loaded with large stinging cells.

together for the good of the organism. They are however much more organised: they share a communal gut cavity and their activities are coordinated by means of a continuous nerve net. But because of this they lack much of the sponge's ability to regenerate—the complex siphonophores are almost incapable of replacing lost parts (other than the stinging cells).

The cnidarians are looked upon as the nearest living things to the ancestral stock that gave rise to all higher animals. They consist of only two layers of cells: an outer, the ectoderm, and an inner, the endoderm. Experiment has shown that these are not interchangeable and that, in those members of the group that do have powers of regeneration, the cells from neither the outer nor the inner layers can alone produce a new individual. More than all other groups of organisms, it is the complex polymorphic forms of the cnidarians that make me believe in the idea of experiments in evolution. Here is a group that, lacking the sophistication of organs to perform specific functions, has instead a whole range of individual polyps and medusae that share the tasks of living. The jellyfish, hydroids, millepores, anemones, corals, and bell-buoys are all living experiments, variations on a theme of success.

Squirrel fish among gorgonians.

An exquisite trade mark
comb jellies

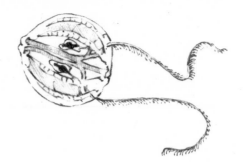

One other group rivals the jellyfish in all their transparent beauty, and although they are not as common they can be found in great abundance in certain waters. Their name is the *Ctenophora*, or comb jellies. They too are riders of the surface range, and like the jellyfish they trawl the wealth of pelagic life.

To a casual observer they would indeed pass for jellyfish, but closer inspection reveals their most exquisite trade mark—eight rows of scintillating plates that flicker with a blue iridescence. These are the comb rows and each tooth in each comb consists of long cilia fused together to form plates that flicker in an ordered wave running down the length of the body. Apart from the beautiful combs, one other welcome feature of these animals distinguishes them from the *Cnidaria*, and that is their complete lack of stinging cells. Not being polymorphic, they are much easier organisms to study—a sea gooseberry, one of the commoner comb jellies, always looks like a sea gooseberry (apart from a short phase when it exists as a larva). Furthermore they exhibit no complex alternation of generations and they never form colonies.

Like the jellyfish, their bodies are made up of two layers of cells, one on the outside, the other lining the internal cavity, with a stiff jelly separating the two. However the jelly of the ctenophores contains many cells from which the muscle fibres are formed. Their internal digestive system is complex: it consists of chambers and passages radiating out from the mouth into the jelly that composes the bulk of the animal, and the gut terminates in long canals running down the length of the body just below the comb rows. The most advanced feature of the gut is that it is, at least in part, a one-way system. Food passes in through the mouth at one end of the body and any indigestible material is passed out at the other end via the anal pores. Digestion takes place partly in the cavity of the gut and is completed after the particles have been engulfed by the cells lining it. Complex it may be, but the gut cannot be regarded as an organ; it is simply an internal cavity lined with special cells whose main function is the digestion of food.

As in the cnidarians, the gut serves another function—to keep the living tissues supplied with oxygen. The correct name for this internal complex is thus the gastrovascular cavity, for it serves a dual role which in more complex animals is split between two organ systems, the gut and the blood or vascular system. The interesting thing is that in both these groups the jelly forming the bulk of the animal is non-living and therefore once made does not require oxygen or food. The tissues that line the outside and inside, and the muscles and other cells found in the jelly itself, must however be kept supplied with these essentials of life. In the animals of the organ grade of construction, the jelly is replaced by living tissue and thus arose the problems of the nutrition and oxygenation of the totally living body that necessitated greater complexities of organisation.

The comb jellies have a basic radial symmetry. This is modified into a tetramerous form by the shape and ramifications of the gut and presence of the comb rows, and further modified into a biradial symmetry due to the positioning of

Sea gooseberry with tentacles 'reeled' in.

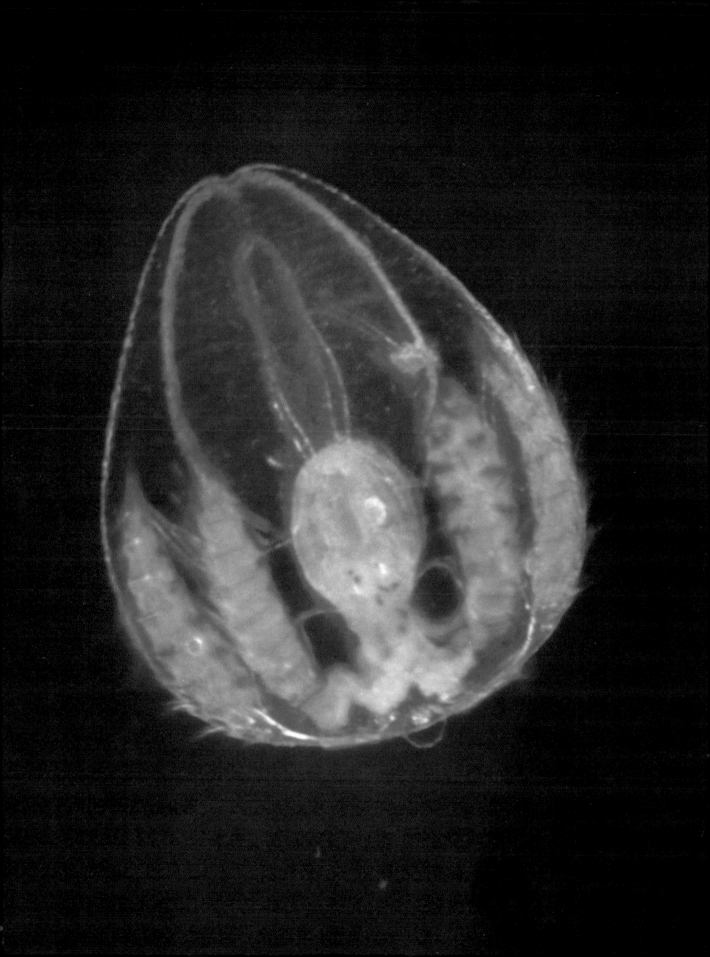

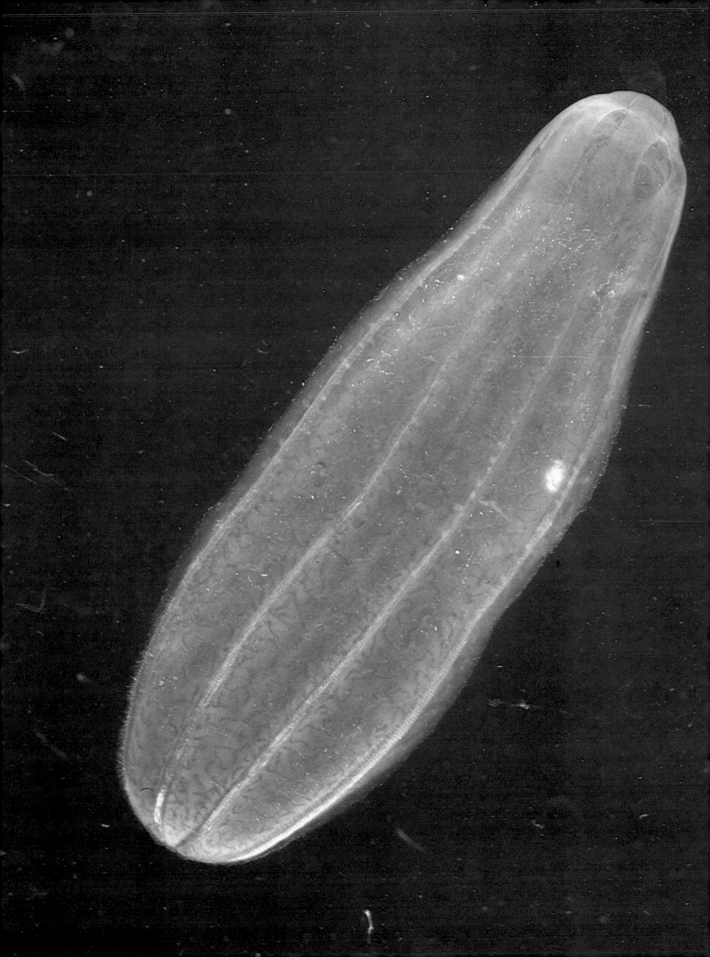

two, long, branched tentacles that may be retracted into pockets, and also in some groups due to the marked flattening of the body.

The tentacles are armed with knob-like batteries of adhesive cells. Although they are not as deadly as the stinging threads of the jellyfish, they are just as efficient. Their movements are also much more controlled in that they can position themselves quite accurately, partly by being held on the surface film and partly by the flickering comb rows. Gentle stimulus of one end of a comb jelly would lead to an immediate reversal of the comb rows and a smart exit (at least for a comb jelly).

The main sensory region is at the opposite end to the mouth where, between the openings of the four anal ducts, there is a very complex organ of balance housed in a transparent dome. It consists of a central mass of chalk granules suspended in the centre of the cavity by four balancing threads made of fused cilia. The whole area and especially the walls of the cavity are well supplied with nerve cells that connect to the nerve net extending to every part of the animal. Special tracks of nerve cells also run down the length of the body just below the comb rows, which are themselves directly connected to the balancing organ by means of four

ciliated channels. It is therefore easy to understand how every flicker of the comb plate is transmitted to the sensory region and the whole movement of the animal is accurately controlled.

Out of all the groups of animals, it would appear that evolution has done most to disguise the identity of some of the ctenophores. Taking the sea gooseberry as a prime example, it is difficult to link its structure to that of the Venus girdle, which grows to a length of nearly two metres and yet is only a few millimetres thick. Only close inspection shows up the tell-tale comb rows, thus clinching the identity of this lady of the warmer seas. The Venus girdle is like a sea gooseberry that has been pulled out sideways, and for this reason four of the comb rows are very short, and four are very long, running the entire length, or rather width, of the upper edge of the girdle. The main retractile tentacles are reduced to two short tufts, but their function has been taken over by a series of short tentacles fringing the lower edge of the transparent body.

The commonest of all the comb jellies, or at least the commonest likely to be encountered, is *Beroe* or the moon jelly. *Beroe*, which can grow to a height of 20 centimetres, is often suffused by a delicate pink colour and can be found in its

A Venus girdle with its tell-tale comb rows.

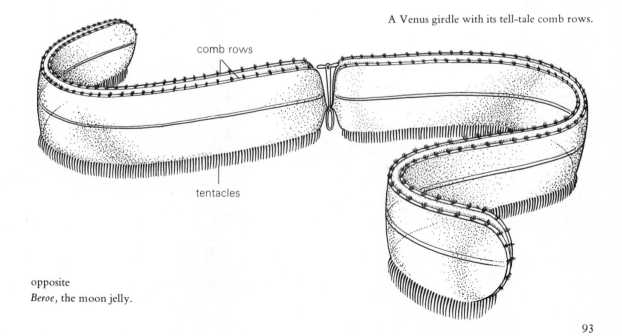

comb rows

tentacles

opposite
Beroe, the moon jelly.

93

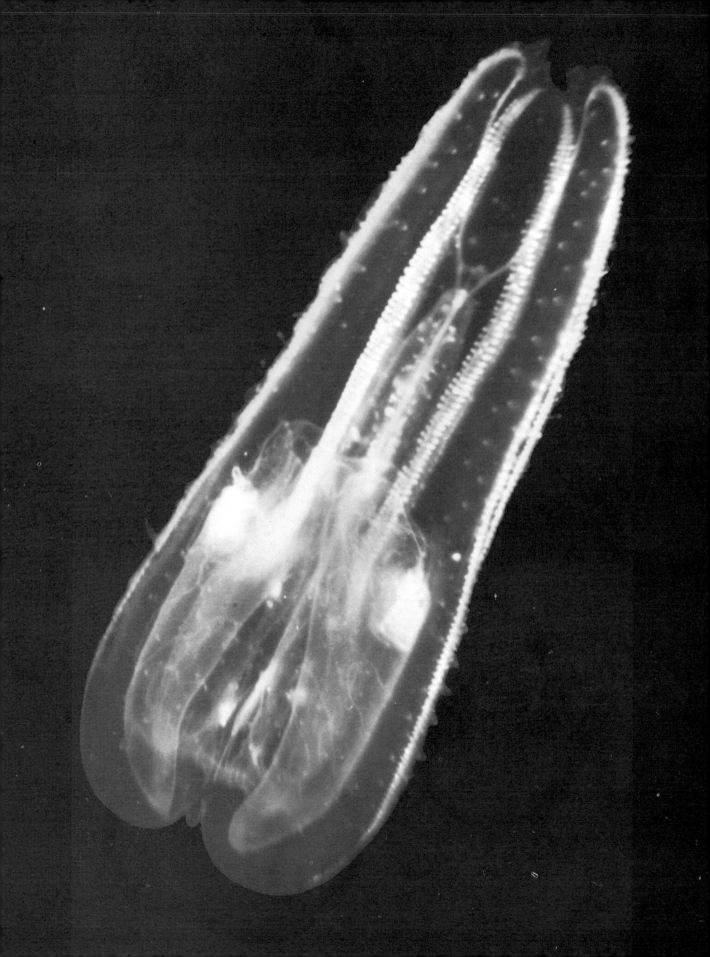

tens of thousands bottled up in bays and fjords by onshore winds. Should you happen upon either sea gooseberries or moon jellies recently washed on shore, it is well worth collecting some and keeping them in a bottle of sea water so that you can take a really close look. If you are lucky and have a well-aerated marine aquarium then it is possible to keep them as the most unusual and decorative of pets.

Like all the comb jellies, *Beroe* is a carnivore that feeds on other animals even to the extent of cannibalism. In Roland von Hentig's classic films about life in the North Sea, there is a sequence in which a sea gooseberry is seen catching and eating a small crab larva. Having started to digest the meal the gooseberry is engulfed by a large moon jelly. The final shot shows the complete food chain in gory detail: inside the moon jelly is a rapidly disintegrating gooseberry, its comb rows still flickering; inside the gooseberry are the remains of the much tougher crab larva; and inside this can still be seen the larva's own gut full of pigmented phytoplankton. With the sun in the background that one beautiful picture has caught with perfection the complete intricate web of life.

The broad, thin Venus girdle may at first sight be difficult to relate to the comb jellies but she is an easy customer compared to some of the others, especially the very small ones. However, in most cases the presence of the comb rows is sufficient to give the evolutionary game away.

In 1886 a zoologist by the name of Korotneff found a peculiar animal in the sea off Sumatra. It puzzled him greatly but he described it as well as he could, calling it *Ctenoplana*. Seven years later four more were found floating on the bone of a cuttlefish off New Guinea. Each was about six millimetres long and of a drab, mottled olive-brown and dull red colour. The body was flattened, with a long, branched, retractable tentacle at either end, and at the centre of the top of the animal was a complex balancing organ connected by ciliated channels to eight comb rows. Undoubtedly *Ctenoplana* was a comb jelly; in fact it was the first to be described in a brand new group of the ctenophores. One of the most exciting things about *Ctenoplana* was that its reproductive cells lie in special cavities that open to the surface by means of a pore. The question was therefore raised: can these in fact be called organs of reproduction?

The next member of this group was found in the Red Sea, where it grows to a length of about six centimetres and lives a sessile existence, creeping about on the outside of colonies of a specific type of dead men's fingers. At first glance *Coeloplana*, as it was called, is not unlike *Ctenoplana* but it lacked the distinguishing rows of combs. Not to be outdone, a further member of the group soon turned up in Greenland's cold and icy waters and it was christened with the impossible name of *Tjalfiella*, the reason perhaps being that it is an impossible animal.

opposite
Bolinopsis infundibulum, plastic poetry in a bag full of motion.

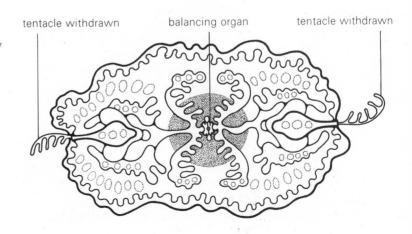

tentacle withdrawn   balancing organ   tentacle withdrawn

Coeloplana—flatworm? no, undoubtedly a comb jelly.

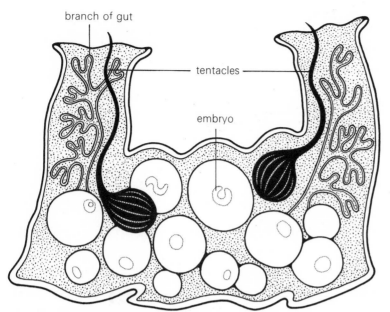

branch of gut

tentacles

embryo

Tjalfiella, another comb jelly in disguise.

Tjalfiella is flattened and each end is furnished with a simple tentacle that protrudes from a chimney-like structure formed from the up-turned, folded ends of its body. Apart from the tentacles it bears absolutely no resemblance to a ctenophore and there is no vestige of any comb rows. The ovaries and testicles occur in eight blind sacs that open via a short duct onto the upper surface of the animal. Over each of these is another sac lined with cilia, and these ciliated sacs are thought to be organs of excretion. However, most surprising of all, the eggs develop while still inside the parent, and each grows into a round larva that looks like a miniature goose-berry, with balancing organ, tentacles, and yes, comb rows—absolute irrevocable proof that *Tjalfiella* is a comb jelly. The larvae leave the protection of the brood pouches to swim in among the plankton, but after a very short period they settle down to a life of creeping about on the surface of a sea pen (a sessile cnid-arian) on which they always live.

In some systems of classification the *Cnidaria* and the *Ctenophora* are joined together into one large group or phylum called the *Coelenterata*, which thus contains the sum total of animals of the tissue grade of construction. However the lack of stinging cells and the various advance-ments in their structure, function, and life history are quite sufficient to set the comb jellies aside as a group in their own right.

The long and the thin
worms

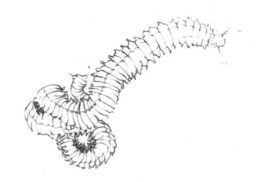

Flatworms

At certain times of the year, along some stretches of the Channel coast of France, the sea shore turns green. Close inspection reveals not a thin covering of seaweed but a seething mass of worms, each one just a few millimetres long looking not unlike tiny green leaves. They are flatworms, members of a very diverse group of animals found throughout the oceans, but being small and almost colourless, most of them are usually passed by unnoticed. Nevertheless, they are important members of the ocean fauna. The green one in question is called *Convoluta roscoffensis* and its colour is due to chlorophyll. However the pigment is not its own, but is contained in the many unicellular plants that inhabit the cells of its tissues.

Flatworms are the simplest of all the animal groups whose structure is based on not two but three distinct layers of cells. Their bodies are 'all alive oh', being devoid of the non-living jelly that makes up the bulk of both the jellyfish and the comb jellies. However, like the cnidarians, the flatworms are of simple structure and they lack any form of blood system (although if they have been feeding on blood they may well become blood coloured until digestion is complete). Perhaps this is one reason why the flatworms are flat—in the absence of blood to carry food and oxygen around the body, the best way to overcome the problem of transport between sides and middle is to remain as slim as possible so that none of the living cells are too far away from the sources of supply.

The gut of the flatworms serves both a respiratory and nutritive function but is more geared to the latter, comprising a muscular pharynx that carries food from the mouth to the highly branched intestine. The majority of flatworms lack any true anal pore although some have developed a one-way digestive system.

The really interesting thing about the green flatworms sunning themselves on the beaches of France is that any food taken into the gut is rapidly digested, and this includes many unicellular plants. The digestive process begins in the 'gut cavity' and is completed after the food has been engulfed by the cells of the gut wall. But not so in the case of the special plants that are to inhabit its cells. These pass unharmed through the whole digestive system and after being engulfed are transported to the tissue where they will live out their lives photosynthesising within the cells of the otherwise white worm. Therefore the plants get a home and in return provide the worm with oxygen.

Experiment has shown that other such 'flat (worm) mates' can keep the worm alive in a corked bottle for over two weeks as long as the light is left switched on; turn off the light and the worms rapidly die from oxygen starvation. Apart from a built-in supply of ready oxygen, there is also evidence that the plants may utilise some of the worm's waste products as raw materials for their own life processes. Thus it is easy to see that a highly complex symbiotic relationship has developed between the two.

It would be safe to say that every major habitat in the oceans has its fair share of flatworms. Apart from the free-living ones they are also found as members of the often complex commensal fauna that inhabits other, larger marine animals. Many sea urchins, sea stars, molluscs, crabs, and fish provide homes for their own specific flatworms, some living on their outsides, others on their insides, but all hitching a ride and enjoying more sheltered lives than their free-living relatives. These commensal flatworms are often modified to their particular mode of life. They lack the conspicuous eyes so typical of the group, and each has either adhesive areas or organs on the surface of their bodies. Perhaps most ominous of all, many have much enhanced reproductive capabilities and are able to produce a surfeit of eggs—a trade mark of the real parasites. Parasites are animals or plants that not only make their home in other animals but derive their food from either the food or the actual tissues of their hosts; and the flatworms have gone in for parasitism in a big way. This group includes such nasties as the flukes and tapeworms, the longest of which, the giant fish tapeworm, can grow to the incredible length of 20 metres inside the intestine of human beings.

The head of most tapeworms is very small in comparison to its body. It is modified for hanging onto the gut wall, and the range of hooks and suckers it uses is a real nightmare. The rest of the tapeworm is no more than an elongate egg factory: its main body consists of flat segments (and there may be as many as 4000 in a big one) each of which is stuffed full of eggs. Living as it does bathed in someone else's pre-digested food, the worm has no need of its own digestive system, and this means more space for the eggs.

The number of eggs produced is prodigious but it needs to be because the road to the next unfortunate human host is a long and complex one, fraught with all sorts of dangers for the developing egg. In the case of the giant tapeworm, the egg passes out with the faeces of its host and develops into a round larva protected in a double membrane. This must then be eaten by a small shrimp-like creature but it must be one of the right sort, because the wrong sort will simply digest it. Once safely inside its new home the larva changes form and bores its way

out of the stomach and into the body cavity of its unfortunate host. Here it undergoes a further structural change, and then remains there until the 'shrimp' is eaten by a fish. Again burrowing from the gut (and one presumes causing both pain and damage as it goes), it settles down to a resting stage in the muscles or other internal organs of the fish. The life cycle is now almost complete. If man then comes along and eats the fish either raw or only partially cooked, he is in for trouble, really long trouble.

The chances of any single egg getting to a new human host are infinitesimally small, which must be good news for us. However, evolution tries to overcome that problem. The tapeworm spends most of its life, and a lot of energy, producing millions of eggs, and as it is our energy in which it is investing, it must surely be on to a good thing. Interestingly enough, as man evolved socially, he partly overcame the problem by eating only well-cooked meat. However now, in these days of uncooked cocktail delicacies, the incidence of tapeworms in highly sophisticated societies is on the increase, and the trouble is that a human being of the cocktail party variety is often so well fed that he or she may harbour a tapeworm for a long time before showing any signs of an inside problem. Life assurance in tapeworms is thus based on very heavy investment in reproduction. Like most flatworms the tapeworms are hermaphrodite, that is, each one has both male and female organs of reproduction, thus they side-step the enormous odds against getting two worms of the opposite sex into the same host.

There are two really fascinating things about the chancy life of the parasites. The first is how these complex life histories, which entail intra-relationships with members of two entirely different animal groups, ever evolved. The second is the remarkable malleability of expression of the genetic information. At each stage in the life cycle it is the same genetic message, originally contained in the nucleus of the fertilised egg, that is being passed on. Yet during each phase this information is translated into an entirely different form, fitted to the life in that particular habitat.

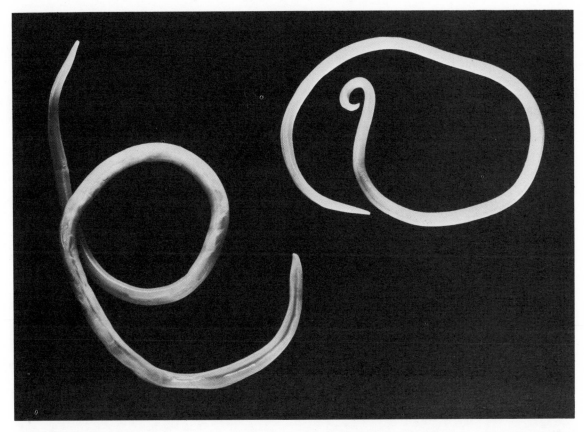

Basic plan of success, a nematode.

The great tapeworm is in fact a parasitic denizen of fresh water fish and is therefore really outside the field of this book. It has been included for two reasons: the morbid curiosity that most people have regarding such things, and that being such a menace to western society it has been studied in great detail. There are however many other parasitic tapeworms which can affect man that find their hosts in marine fish and even in seals and whales.

Not all flatworms are such nasty, revolting creatures and many of those that live a free life in the warm tropical waters are extremely beautiful, adding bright jewels of slowly moving colour to the reefs. Others live a planktonic existence partly stuck onto the surface film of water, moving by means of their cilia-covered outer tissue. All the free-living forms move in this way on thousands of mini legs, so that their progression is a slow and graceful glide with hardly a movement of the actual body of the worm.

There is another unusual feature of the digestive system of these truly remarkable animals and that is that some can feed on members of the *Cnidaria*, digesting their soft parts but leaving the stinging cells intact. The stinging cells are enveloped by the cells lining the flatworm's stomach and are then passed out to their 'skin' where they are used for defence and food capture. There is also some evidence that these flatworms only feed on cnidarians when they need a new supply of stinging cells. If substantiated, such behaviour would point not only to a high degree of nervous control of the flatworm's activities but would also indicate a certain degree of learning.

Apart from their complex system of reproductive organs the most advanced feature of the flatworms is the ordered arrangement of their nervous system. At the head end definite masses of nervous tissue may be seen, from which tracks of nerve cells run down the length of the body joining up as they go with the general nerve network. Much work has been carried out showing their nervous system to be quite sophisticated. They react to light, food, currents, chemicals, and electric shocks. Experiments with *Stenostomum*, a flatworm that inhabits shady places in the sea, show that it has a certain capacity for learning. It can be trained to turn back from what would be its normal light environment by punishing it with mild electric shocks, thus giving undoubted proof of associated learning, although the retention time is very short. The seat of the short memory would also appear to be in the nerve ganglia because their removal produces an untrainable flatworm.

If experimental flatworms can live minus their 'brains' then it is hardly surprising that they are able to regenerate lost parts; and they certainly do have remarkable powers in this field. However in those possessing a definite head end it is the part with the brain that usually regenerates the best, the rest of the body taking a long time to grow a new, functional head.

Proboscis worms

Of all the monsters of the deep at least one is not mythical; *Lineus longissimus* has been seen in and collected from the North Sea. This is the longest of the proboscis worms, which are best described as animated bootlaces, for many of them are very thin in relation to their length. Libbie Hyman writes of *Lineus* the longest in guarded tones: 'it is said to attain a length of 30 metres'. If it does then it certainly must be a fantastic creature because the feature shared by all the members of this group, called *Nemertea*,

A ragworm at speed. Note the snake-like motion and the action of the 'paddles'.

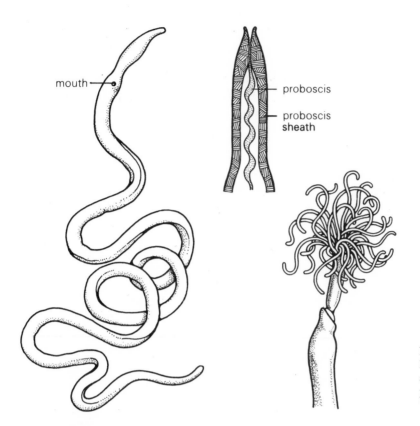

mouth

proboscis

proboscis sheath

Proboscis worm with section of its head to show the wondrous nose, and *Gorgonorhynchus* with its proboscis unfurled—the original 'Afro' haircut.

is that they have a proboscis that is usually two to three times their body length. This mighty nose normally lies in a special cavity that occupies much of the worm's body. It can be shot out very rapidly, everting as it goes, and this would mean that, when fully extended, *Lineus longissimus* could be over 90 metres long.

Ever since I took up diving I have searched the North Sea in vain for what must be the longest, thinnest creature in the world. To date my most exciting discovery has been a 12 metre length of brown bootlace seaweed, *Corda filum*, which I approached in trepidation, waiting in vain for that nose of vast proportions to shoot ·out at me.

Most of the proboscis worms are much shorter—a fair sized one would measure about 20 centimetres. They creep about in the nooks and crannies mainly below low water mark, and although some of them are brightly coloured they are difficult things to find. The effort is however well spent, especially if you are

fortunate enough to see the proboscis in action. It flicks out with remarkable speed and the end, which may be armed with sharp, penetrating stylets, wraps itself around the prey. More often than not their prey is another worm, for it is not difficult to understand that a bootlace-shaped organism would have terrible digestive problems if it tried to swallow anything not of similar design. The proboscis has many other functions. When in a hurry, what could be easier than a long reach forward, a grab hold of any solid object, and pull?—a real case of 'nosing about'.

Not all of them live in natural crevices: some dig burrows, and here the nose forms an excellent 'hydraulic ram' excavator. Others with a more flattened design lead a pelagic life, swimming in mid water where the long everted proboscis aids both floatation and propulsion. In any family there is usually at least one zany member, and here it is *Gorgonorhynchus*. This has a proboscis with a tuft of smaller probosces at the end looking rather like an 'Afro' haircut, but ideal for catching lots of worms. The most remarkable part of the wonderful proboscis is the retractor muscle which allows it to be pulled in once it has been shot out. When contracted it may be only a few centimetres long but extended . . . well in *Lineus longissimus* the mind boggles.

Apart from their distinctive noses, the nemerteans are just simple worms, although they do show a number of advanced characteristics when compared to their flat cousins. Their food goes in at one end and the indigestible part comes out at the other. Their skin is complex, consisting of an outer layer of ciliated cells among and below which are numerous gland cells that secrete copious amounts of mucus, thereby laying a slimy trail along which the worm crawls. Underneath the surface layer is a definite dermis and below this is a complex sheath of muscle, whose outer part encircles the worm, while the inner part runs down its length.

The whole body is well supplied with blood vessels, some of which are contractile and pump the blood around the complex of tubes. At certain points special cells, which are connected via ducts to pores opening on the outside of the worm, lie close against the blood system. The function of this pore system is excretion, that is, the removal of the toxic by-products of the life process from the tissues of the worm. The nervous system is likewise more complex and a high level of muscular co-ordination is evident from the complex swimming movements of the worms living a pelagic life.

Sophisticated animals, well advanced along the organ grade of construction, they may be, and yet their powers of regeneration are amazing. Simple fragmentation of the body leads to multiplication. Each fragment, containing the full complement of tissues, is able to grow into a new worm, nose and all.

Most remarkable is the capability of certain of the proboscis worms to overcome very long periods of starvation. Experiments have been continued for as long as two years, during which time the worm not only shrinks but actually returns almost to its embryonic condition. At first the cells of the gut start to digest their neighbours. Then the cells of the body become amoeboid and wander through the worm, engulfing whatever comes their way. Finally complete de-differentiation takes place, with the emaciated animal ending up as no more than an ovoid body comprising an inner and an outer layer with loose wandering cells in between. But given time and a fresh supply of food, the whole process can be reversed until the worm, like Edward Lear's famed Dong complete with a 'wondrous nose of vast proportions', sallies forth once again.

Pseudocoelomates

It is among the true or segmented worms that we see the result of the next major breakthrough in evolution, and one that set the pattern of all the more complex animals.

The true worms are coelomates, which means that much of their bulk is occupied by dead space. Apart from their internal 'gut', jellyfish are solid jelly, flatworms solid flatworm, and the body of the proboscis worms solid proboscis worm (except for the cavity to house their very special noses). But the true worms are without doubt hollow. In structure they consist of a body wall of skin and muscle inside which is a space, and suspended in this space is a one-way gut, together with the other organ systems. All the living tissues are supplied with the ramifications of both the blood and nervous systems, which bridge the gap in the tissue extensions that suspend and hold the gut in position.

The space is called the coelom or body cavity. It is formed during the development of the embryo, and is completely lined with a special tissue called the peritoneum. Like the jelly of the jellyfish, the coelom has the advantage of adding bulk without requiring massive and continuous supplies of food and oxygen for its upkeep. It also allows room for the development of the complex of organ systems. With a folded muscular gut to deal with tougher types of food, and complicated and efficient excretory and reproductive systems, worms are living

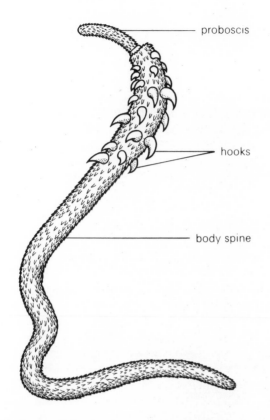

proboscis

hooks

body spine

The hooks of a hook worm, aptly named *Rhabdo-rhynchus horridus*.

proof that the possession of a body cavity overcame some of the limitations of size.

There are however three groups of worms that may conveniently be linked under the term pseudocoelomates or false coelomates, of which at least some, apart from being important experiments in evolution, must be reckoned among the most successful of the animal kingdom.

The first group are only likely to be met by fishermen and wildfowlers when cleaning their catches, for they are all parasitic and live the main part of their lives in the gut of fish and fowl. They are repulsive creatures when regarded with human sensibilities, and parts of the unfortunate host's inside may be crammed full of worms. Each hangs on to the gut wall by means of a short evertible proboscis furnished with an array of hooks, hence their common

name, the hook worms. Like all true parasites their reproductive capacity is prodigious to say the least. In one species it has been calculated that potential offspring are produced at the rate of 260,000 per day for a period of ten months—which makes an incredible total of over 79 million! The feature linking the horrible hook worms to the other members of this diverse group is their possession of a false coelom. Their body cavity is not lined by a special tissue, but is simply a gap that develops between tissues.

My own real introduction to the next group, the truly phenomenal rotifers, was at a meeting of the Royal Microscopical Society in London. I had been taken by my father, a keen microscopist, to hear a lecture by Professor Jane on the structure of wood. Fascinating as his talk was, my attention was caught by the row of large books adorning the wall against which I sat—thirty-seven tomes or more on the *Rotifera*.

The crowning glory of the rotifers is a circlet or circlets of long cilia that surround the mouth. When in motion the regular beating of these cilia looks like a rotating wheel pulling them through the water, hence their common name, the little wheel animals. Being so small (they range from 0.04 to 2 millimetres in length), for a long time they were reckoned to be unicellular protists. The game is given away by a very nasty pair of grasping jaws which are shot out from the middle of the wheel to deal effectively with the next meal. The jaws surround a mouth that is connected to a complex one-way gut; this opens via an anus, also located within the halo of cilia. In addition the body contains reproductive organs, a nervous system, and a definite false cavity; thus rotifers are complex organ grade organisms and what they lack in size they certainly make up for in number. There are at least 15,000 different species although, surprisingly enough, very few are exclusively creatures of the sea. The best place to see them is among the grains of beach sand (three cubic centimetres of water from such sand can contain as many as 400 rotifers), but of course you would need a microscope.

When it comes to sheer numbers there is another group, the nematodes, that puts the rotifers, and come to that all the other groups of many celled animals, to shame. Ralph Buchsbaum, a well-known zoologist, writing of the nematodes or round worms said that if all other plants and animals were wiped off the face of the earth, you could still see where each and every individual had been by the nematodes it left behind. So amazing is this statement—and so fantastic is this group of animals—that it deserves explanation.

Of this same group, Libbie Hyman said: 'It was long ago pointed out by Cobb that every vertebrate is infested with at least one and usually with more than one kind of nematode and as there are some 45,000 species of vertebrates it is clear that at least 100,000 nematode parasites of vertebrates must exist. To these must be added the phytoparasitic forms and those that parasitise invertebrates, chiefly molluscs, crustaceans, insects, centipedes, and millipedes. Formidable as this array of parasitic nematodes is, the figures become more incredible when it is considered that the free living nematodes outnumber in species the parasitic ones. It is therefore a reasonable assumption that there are at least 500,000 different sorts of nematodes in the world.'

They are everywhere, in enormous numbers —yes, even inside you and me. However the majority are so small that we can neither see nor feel them moving about, but under the microscope it is a different matter. They all have the characteristic eel worm shape—long, thin, and tapering almost to a point at either end. Their movements give them away, and give them their name too, the eel worms. They thrash about using their whole body in a very violent eel-like motion that gets them nowhere fast.

Revolting as it may seem to be surrounded by, and to actually surround, many writhing worms, it is a fact that these tiny creatures are of great importance to the economy of nature, for many of them live on the products of decay and so help clean up the environment. Although the average visitor to the sea will undoubtedly come into contact with nematodes, he will never see them.

Fishermen may find parts of their catch infested with some of the larger parasitic worms. The longest can reach a horrible 40 centimetres, and can be dissected to reveal the large, false body cavity devoid of peritoneum. These are the biggest of the pseudocoelomates and it would appear that some factor, which we do not yet understand, limited their further development. From here on upwards in the evolution of complexity, the true coelom with its delicate lining was the 'in thing', in more ways than one.

Segmented (true) worms

There is a saying, 'nothing is impossible', but have you ever tried standing a worm on end? The reason it cannot be done is that worms do not have skeletons, at least not in the accepted sense. However they do have true body cavities which are full of a special fluid, and this is of great importance because it gives the body some measure of rigidity against which the muscles lining the body wall can act. The coelom, together with its coelomic fluid, thus forms a hydrostatic skeleton, ideal for wriggling but no good for standing up.

The majority of the true worms are segmented and each constriction on the outside of the animal marks the position of an internal septum. These effectively divide the long body cavity into a number of compartments, each a separate hydraulic unit, and each adding a little bit of rigidity. Only the true worms are segmented; the other worms that appear segmented from the outside lack the internal divisions.

If you inadvertently cut an earthworm in half while digging in the garden, very little coelomic fluid is lost and the cut end is rapidly closed off by constriction of the circular muscles of the body wall. The head end goes away a little slower than before and probably a little wiser than the tail end, which is no longer connected to a brain.

The underside of a garden worm feels rough to the touch. This is due to stiff bristles or chaetea, four of which adorn each segment and help the animal to get a grip while moving along.

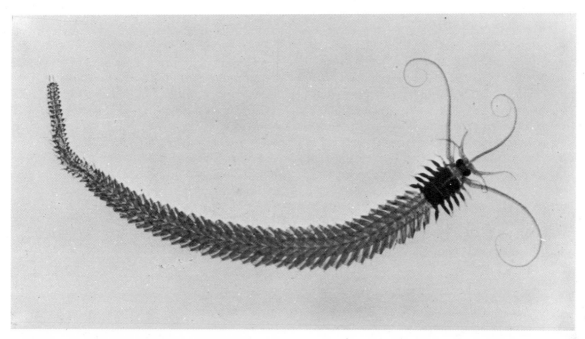

Autolytus, an errant (free swimming) polychaete or
paddle worm. How many segments are there?

Head of a polychaete *Perinereis cultrifera* showing the
formidable jaws and the body segments.

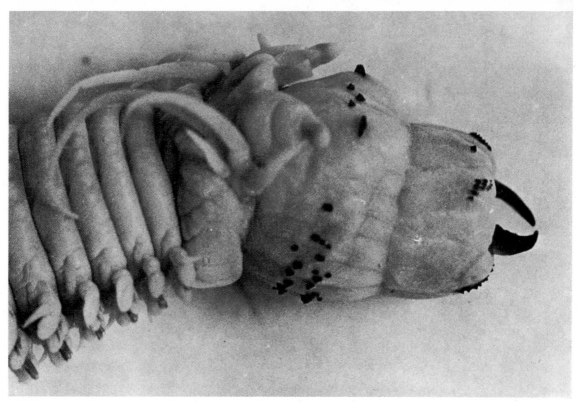

Garden worms and their kin have few bristles and are therefore called oligochaetes, the Greek word *oligos* meaning few. Many of the marine worms are much more flamboyant things. They have many bristles (hence their name polychaetes) which are borne on the tips of what look like stumpy legs, the parapodia, one on either side of each segment. The polychaetes come in three basic forms, each related to their mode of life: some are errant (they live in part a free life, swimming through the water), some dig burrows, while others build the most fantastic tubes in which they happily live their lives away.

The palolo worm, like all errant worms, has a head with a short proboscis armed with a pair of jaws. It moves through the water by an undulation of its whole body, rather like the sea monsters of horror movies, and it lives in holes in the rock from where it makes brief sorties to find food. On one particular day in each year the palolo, or at least part of it, is destined to make a very special journey. As in most marine worms, the eggs and sperm pass into the coelom before being released by special ducts out into the sea where fertilisation takes place.

The parchment worm *Chaetopterus variopedatus* living up to its second name, which means many sorts of legs. The horseshoe-shaped structure holds the mucus trap. Note the food mass and the groove running forward to the mouth.

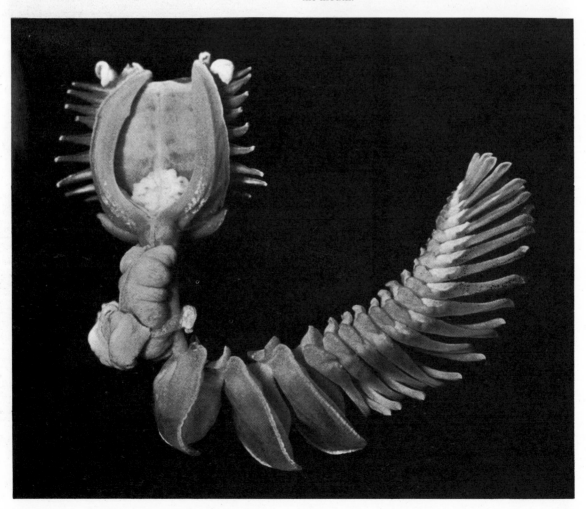

A tube full of beauty. The peacock worm *Sabella pavonina* extending its tentacles.

In the southern Pacific, this occurs during the last quarter of the October-November moon: the rear half of the palolo worm fills with eggs and sperm, and when the day of the last quarter dawns the worm sticks its ripe tail end out of its home and sets it free into the ocean. The errant half rises to the surface where the eggs and sperm are released in such numbers that the sea turns milky. The front end sits tight in its coral home and gets on with the job of producing a new tail end in readiness for exactly the same dawn operation the following year. The palolo of the western Indian Ocean repeats its annual performance in the third quarter of the June-July moon. How these creatures are exactly plugged in to the phases of the moon is not known, but it is such a regular and reliable occurrence that the locals set their dietary clocks by it each year, coming to gorge themselves on the worm 'roes' that can be scooped up out of the milky water.

If the prize for precision timing goes to the palolo, then the prize for the most bizarre modifications of the bristle-bearing parapodia must go to *Chaetopterus*, a weird looking worm that builds a fully detached parchment-like tube. *Chaetopterus* is an accommodating worm for, apart from sharing its tube with a number of commensal flatworms, it can be removed and will live quite happily in a U-shaped glass tube open to inspection. The tube it lives in must be of the right calibre for the worm to fit in loosely, allowing sufficient room for the rhythmic flapping of the fans of the middle region of the body to draw a stream of oxygenated sea water through it. With the water comes food, which is caught in a mucus bag held out in the water stream. At the base of the bag is a small cup in which the food is moulded into a ball. When this ball becomes the correct size it is passed forward along a special ciliated groove to be popped into the mouth—complicated, but very effective.

The tube dwellers that delight all skin divers are the peacock worms, whose great fan-like masses of gorgeously coloured tentacles protrude from the mouth of the tube. On one side of each tentacle are special, long cilia that vibrate in unison to carry particles down into a groove leading to the mouth at the base of the crown of tentacles. Before they reach the mouth they have to pass through a sorting device consisting of a deep, tapering, trough-like groove. The smallest particles can fall to the

Christmas tree worm with its tentacles ready for a meal.

opposite
Dalyellia, a flatworm living on borrowed chlorophyll.

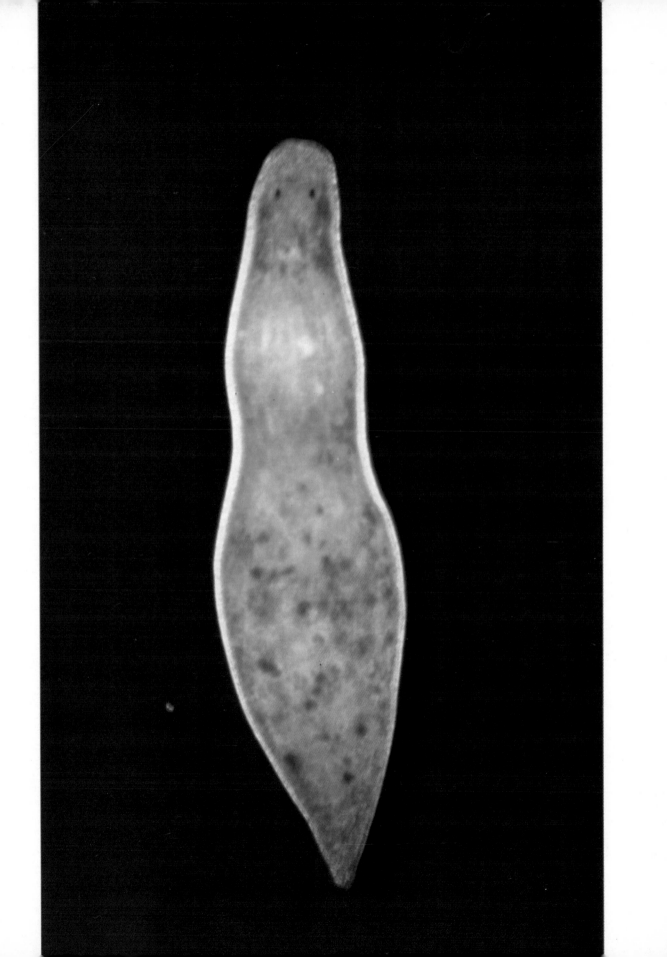

bottom of the trough and are passed to the mouth. Medium sized particles do not fall in as far, as they are passed backwards to be used for tube construction. The largest particles simply pass along the top of the groove to be rejected. To see the wonderful tentacular crown at work, the worm must be approached very carefully because at the slightest sense of danger the peacock withdraws in a flash, leaving a rather drab, dead-looking tube.

A walk on a muddy shore at low tide will usually allow you to see, if not the burrowing worms themselves, at least the results of their activities—a neat cone of mud and not far away an equally neat cone-like depression. Between these two lives a lug worm. It lacks the great crown of tentacles and its short proboscis is armed only with small warty papillae, for this is both its main organ for burrowing through the silt and for feeding. Life is all ups and downs for the lug worm. It crawls forwards and upwards, ingesting silt as it goes and hence causing the

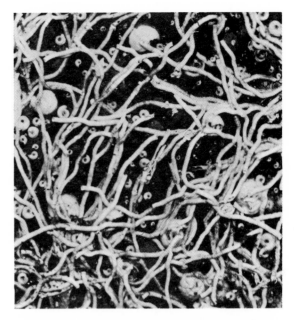

A mess of worm tubes, all made of chalk. Each of the coils contains a *Spirorbis*; the straighter ones belong to *Filograna implexa*.

opposite
A flatworm goes about its business.

Tubes of the sand mason worm
Lanice conchilega.

depression. Any organic matter contained in the silt is digested, and the inorganic particles of sand and silt pass straight through and out at the other end. When full the lug worm reverses and crawls backwards up the other arm of its U-shaped burrow, depositing the tell-tale worm casts in a pile over its back door.

The true worms belong to the organ grade of construction, and they are capable of extremely organised life patterns. Their extensive body cavities are used both for the storage of the gonads, the eggs, the sperm, and also the waste products. If their waste products were not got rid of they would poison the body, and a feature of the worms is their complex organs of excretion which gather up the toxic substances and pass them to the outside. These organs are called nephridia and are the forerunners of the kidneys of the more complex animals.

Vampire 33

A good composite title for the last section on the true worms would have been 'The importance of being segmented' or even 'Too many of a good thing'! This would have given a nice lead into the following chapter on the crustaceans and formed a meaningful link between the two great groups of obviously segmented animals, the annelids or worms and the arthropods.

There is, however, one small group of animals that sets up this link quite naturally and they are the nasty, horrible, revolting, but really rather interesting blood-sucking leeches. Leeches are undoubtedly segmented worms, but they do have a number of peculiar characteristics that link these, the most malleable, ductile, and inflatable of all animals, with the highly honourable armoured corps of crabs and insects.

Their most constant feature is their number of body segments, thirty-three to be exact. Looking at the average leech there would appear to be many more, but the number of outer markings does not relate to the internal segments. The only clear external evidence of the countdown of true segmentation is the ordered rows of sensillae, which are minute, white, dot-like sense organs found on most specimens. The leeches have far fewer segments than the worms and there is little doubt that this reduction in number has helped to increase both the mobility and the agility of the blood suckers, a feature that must be of great importance to an animal whose food source has to be

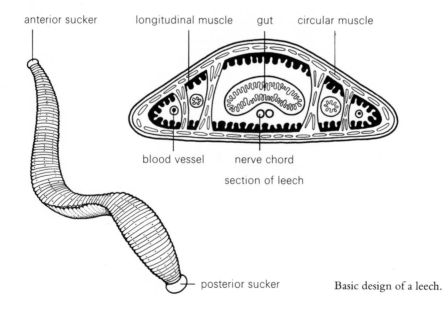

anterior sucker   longitudinal muscle   gut   circular muscle

blood vessel   nerve chord

section of leech

posterior sucker

Basic design of a leech.

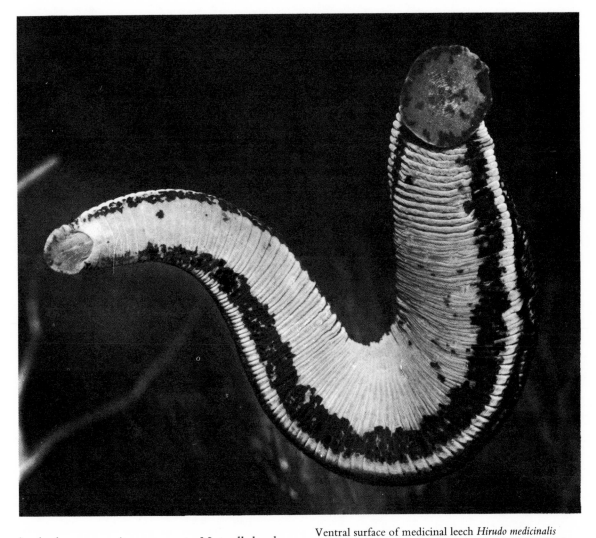

Ventral surface of medicinal leech *Hirudo medicinalis* showing segmentation and posterior and anterior suckers.

latched onto as it goes past. Not all leeches actually suck blood. Some are active predators, giving chase to their prey (more often than not other worms) and engulfing them with amazing rapidity.

The majority of marine leeches are ectoparasites. Some grab at any passing meal, hitching a ride until, gorged to satisfaction with blood, they fall off, while others live a more permanently attached existence. Their mechanism of attachment consists of two highly efficient suckers, one at either end of the body. The sucker at the front is furnished with a number of pairs of eyes and a muscular proboscis that can be everted to penetrate the victim's skin so

providing direct access to the liquid diet. This group offer little or no threat to the thick-skinned mammals because they lack efficient cutting jaws. But medicinal leeches, which are denizens of fresh water, and some land leeches possess three jaws that act like minute circular saws to inflict the prey, almost painlessly, with the typical Y-shaped leech bite.

All the vampire leeches, whether jawed or proboscised, produce special chemicals inhibiting the clotting of the victim's blood so that the nutritious food flow is not impaired. It then simply pumps out the blood by pulsation of its muscular pharynx. Apart from their nasty feeding habits and pulsating table manners, leeches have many hidden surprises, some of which are even quite endearing. Revolting as they may appear to us, one leech can find another very attractive and many go through definite mating rituals prior to fertilisation, which may be a reciprocal process, because like the tapeworm these animals are hermaphrodite. In the majority of the jawless species the packets of sperm are simply deposited on the surface of the mate from where they are released and, in some way as yet not completely understood, the sperms penetrate the skin and migrate through the body to effect internal fertilisation of the eggs.

Perhaps even more surprising is the fact that a number of them are very good mothers and actually tend the young throughout their early development. The brooding mother can collect as many as fifty developing embryos, which are plugged into her underside by means of special ball-and-socket joints. By gentle undulations of her body she is able to keep them well supplied with oxygenated water for periods of several weeks—devotion indeed.

Leeches are most closely allied to the oligochaetes, the group of worms that includes the garden variety, but they lack the bristles that make the garden worms such rough customers and give them such remarkable powers of traction through the soil. However, their most peculiar adaptation is the complete lack of digestive enzymes in their expandable gut, for with a diet of someone else's blood full of someone else's pre-digested food, who needs them?

It is evident therefore that these animals are worms modified for a semi-parasitic mode of life. But what about the links with the arthropods? Both the arthropods and the leeches have fewer segments than the long, thin worms. The true worms have an extensive fluid-filled coelom that acts as a hydrostatic skeleton; in contrast, the coelom of the crabs and their kin is much reduced, support being provided by the tough exoskeleton, and the blood system takes over the main body cavity. The true coe-

lom of the leeches is almost filled up with packing tissue leaving only a system of tubules through which the coelomic fluid, complete with haemoglobin, circulates to keep the body supplied with oxygen. In many marine leeches this coelomic blood system is continued out into accessory gill-like structures protruding from the body wall.

Leeches thus show an intermediate form between the two groups, but most peculiar of all is the fact that their nervous system, which centres on six pairs of ganglia in the head region, bears at least some resemblance to that of the insects. However, it would be wrong to even suggest that they show an evolutionary link between the worms and arthropods. The similarities with the latter group are probably only fortuitous; nevertheless they are of great interest.

Too many legs
crustaceans

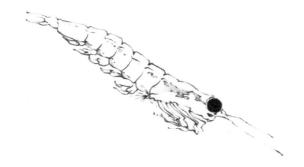

The armoured crabs and lobsters may at first sight seem far removed from the soft bodied worms, but they do have one prominent feature in common—their bodies are segmented and some or all of the segments bear a pair of legs. Apart from the rigid external skeleton, the other feature in which the crustaceans differ from the worms is in their complete lack of cilia. Many of the flatworms are covered with cilia and all the various types of worms have at least some on restricted parts of their bodies. They perform all sorts of functions in locomotion, feeding, reproduction, respiration, and tactile sense, that is, in the sensing of the environment.

In the crustaceans the various functions of the worms' cilia are carried out instead by their wonderful legs, or rather appendages ('leg' sounds as if it were made only for walking). Because of the thick outer skeleton the appendages have to be jointed and certainly the joints appear to be somewhat awkward things. A crab walking sideways and a lobster moving its armoured bulk through the water look ungainly. However a close look at them and especially the highly modified appendages around the mouth will show just how efficient a joint can be. Better still, a walk on a tropical beach in the moonlight would show the speed

Caprella liparioteusis, a fairy shrimp.

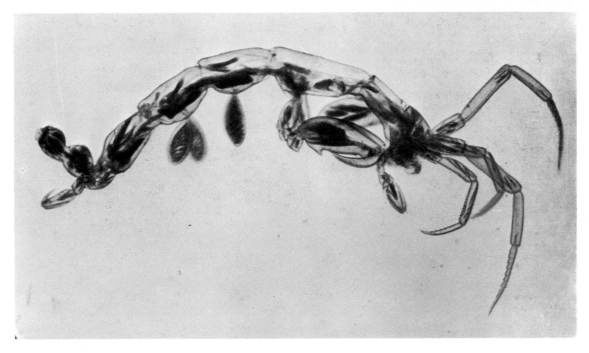

with which the ghost crabs can move and how agile they are as they nip down their burrows.

The crustaceans are just one of the groups of animals that zoologists include within the great phylum of the animal kingdom called the *Arthropoda.* All the arthropods are linked by the bond of external armour, the exoskeleton, and they include among their members the crabs, lobsters, shrimps, and their kin (the *Crustacea*); the centipedes and millipedes (the *Myriapoda,* meaning many feet); the insects (*Insecta*); and the spiders (*Arachnida*).

Of all the phyla, this one contains the largest number of land animals, for practically all insects and spiders dwell on dry land. Perhaps the most surprising fact is that of the million different sorts of insects in existence, none of them actually live in the sea, although a couple do run about on its surface. There is little doubt that it is the hard exoskeleton preventing massive loss of water that opened up the dry earth to this the most successful group of land animals.

However, certain members of the phylum remained tied to the water and many of them to tidal water. They are conveniently termed the crustaceans, the animals with the truly wonderul legs and the heavy armour plating.

Thus the shrimps, prawns, scampi, crayfish, lobsters, and crabs all belong to the same mouth-watering group that live a free life swimming or, more often, walking over the bottom. The legs may be modified to perform one or all of the four basic functions—locomotory, respiratory, feeding, and sensory—and the group as a whole trace the evolution of specialisation of the appendages for these various purposes.

Many of the segments making up the bodies of the fairy shrimps bear lobe-like limbs fringed with bristles that are remarkably similar to the parapodia of the paddle worms. The co-ordinated beating of all the legs drives the shrimp through the water. The same limbs help to collect food particles, and as they are covered with only very thin cuticles they are also the main site for the ingress of oxygen—hence the proper name of the fairy shrimps, the *Branchiopoda,* meaning foot breathers. Furthermore, there is evidence that their delicate appendages sense what is going on in the water round about, and so they have it, the all-purpose foot.

From the branchiopods on, it is specialisation all the way. In certain groups one set of legs is modified to do one job, while in another the equivalent set takes on an entirely different

opposite
'How does that "crab" you!' The underside of a squat lobster with her tail tucked up.

'Come here and I'll show yer.'
World champion *Calappa granulata* from the Canary Islands.

function. This clearly points to the fact that the original stock, from which all the varied crustaceans diverged, possessed legs that were evidently Jacks of all trades, but they were certainly not masters of them all. Thus it would seem that while the basic design of the crustaceans' all-purpose leg was good, at the same time it was ideal for further modification.

Hand in hand with the increased specialisation and efficiency of the limbs, the body became shorter and eventually the number of legs was reduced and segments lost. The losses occurred in the hind region of the body to give the basic

design of decapod (ten-footed) crustaceans like the lobster. The most obvious appendages of the lobster are of course the eight walking legs and the two large vicious looking pincers or chelipeds which, apart from wreaking havoc with any would-be intruder, are also one of the ultimate sea-food delicacies.

It is easy to see, especially when amid lettuce and mayonnaise, that the body of the lobster is made up of three distinct parts. The head, with its stalked eyes that may be hinged sideways back into the protection of the shell, bears five pairs of appendages. The short antennules (little antennae) and the long, long antennae,

Top view of an edible crab *Cancer pagurus*.

The appendages, and their functions, of a lobster.

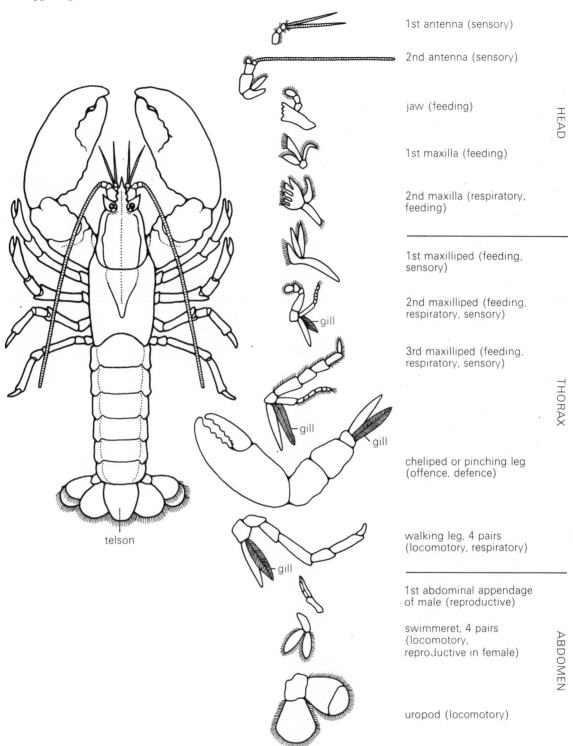

1st antenna (sensory)

2nd antenna (sensory)

jaw (feeding)

1st maxilla (feeding)

2nd maxilla (respiratory, feeding)

HEAD

1st maxilliped (feeding, sensory)

2nd maxilliped (feeding, respiratory, sensory)

3rd maxilliped (feeding, respiratory, sensory)

cheliped or pinching leg (offence, defence)

THORAX

walking leg, 4 pairs (locomotory, respiratory)

1st abdominal appendage of male (reproductive)

swimmeret, 4 pairs (locomotory, reproductive in female)

uropod (locomotory)

ABDOMEN

telson

gill

gill

gill

gill

each mobile from a ball-and-socket joint, are sensory in function. The jaws or mandibles are highly modified for biting, and the maxillae are fan-like and function both in respiration and in helping to pass food to the mouth. The mouth

parts proper are composed of the three sets of maxillipeds (jaw-feet) all modified for feeding, but these are attached to the next portion of the body, the thorax.

The thorax also bears the pincers and the four pairs of walking legs. In the female the reproductive ducts open at the base of the second pair

How many eggs? Underside of a female devil crab *Portunus ruber* in berry (carrying eggs).

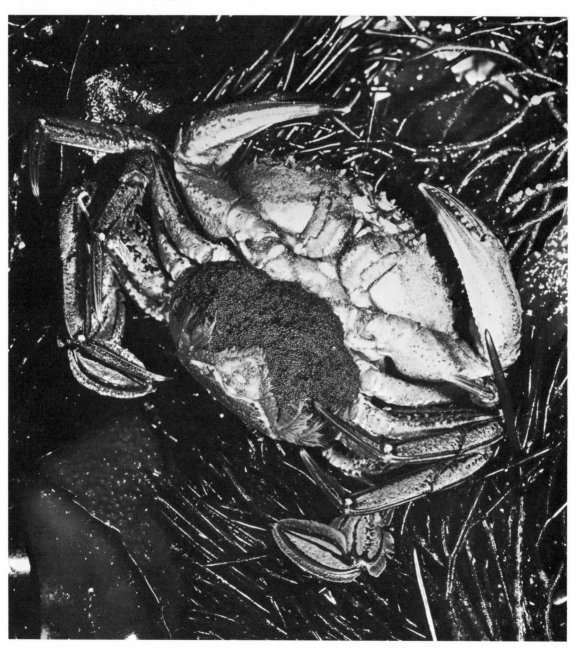

Commensalism or house-sharing, like Easter bonnets, can go too far. *Calliactis parasitica,* a sea anemone, attached to a whelk shell, the fully detached home of *Pagurus bernhardus,* a hermit crab.

of walking legs and in the male at the fourth, thus adding yet another function to the already well-worked legs. Contained in the thorax are many of the vital organs, which are covered and protected by the strong armoured carapace developed from a fold of skin at the back of the head. The shape of the carapace is also of importance as it produces a chamber through which the respiratory currents are maintained, bringing oxygen-rich water to the gills.

Behind the massive thorax is the more delicate but nevertheless immensely strong abdomen, each segment bearing two swimmerets. These appendages are modified both for swimming and for brooding the eggs and protecting the developing young. The abdomen is terminated by the lobes of the broad tail, or telson, beneath which the anus opens, voiding indigestible food material to the outside world. The telson is very important in locomotion, especially when the lobster is on the retreat, for it can fan out to give maximum surface area, and the whole rear end can be flicked forward jack-knifing the lobster backwards through the water.

In contrast, the crabs, and there are very many different sorts ranging from the almost microscopic pea crabs to the enormous spider crabs of Japan measuring more than two metres across, all appear to have lost their abdomens. However the truth is that the abdomen in the crabs is reduced and is turned back underneath the thorax.

One group, the hermit crabs, has retained a relatively large abdomen, but it is an abdomen that presents problems because it has a very thin exoskeleton and is therefore very vulnerable to attack. For this reason the hermit crabs seek the protection of other objects, more often than not the discarded shells of molluscs. However in these days when the sea shore is often littered with man-made objects it is not unusual to find larger specimens with their tender rear ends safe inside the end of a broken bottle that may be glass, china, or even plastic. In spite of this, most stick to natural resources when house hunting, and once inside the safety of a shell the soft part of their bodies is coiled to fit the coils of the shell and one of the claws fits across the entrance like a door. The legs of the abdomen are either missing entirely or are reduced and modified for hanging on to the inside wall of the home.

There is little doubt that it is the weight problem that has restricted the crustaceans to the aquatic medium where the buoyant water bears much of the weight of the massive chalky skeleton. However there is another difficulty associated with a tough exoskeleton—how can the body grow? In the same way that the hermit crabs have set themselves a never ending problem of real estate, so too have the rest of the crustaceans. It is a life-long job, not of keeping up with the Joneses but of keeping ahead of your own vital statistics.

First they must get rid of their old armour, a process not unlike you or me trying to take off our skin, and remember we do not have awkward things like jointed legs and long feelers to get in the way. Once the moult is complete the crustacean is, temporarily, in real trouble for although a new exoskeleton has been formed underneath, it is still soft and therefore of no use whatsoever as a protection against predators. So throughout the procedure the helpless creature, including even the largest of lobsters, must hide in some hole or crevice guarding the entrance as best it can with its soft claws. In approximately thirty days the new shell has grown to size and hardened up sufficiently for the once again tough customer to be free to emerge, ready for anything. Complex as the great decapod crustaceans are they still have exceptional powers of regeneration; even the mightiest of limbs lost in battle may be replaced, enlarging at each moult.

In comparison with the giants of this group, the crabs and the lobsters, the majority of the crustaceans are very small, but despite their size

The larva of a succulent lobster (magnification × 3.8).

these are undoubtedly the most important. Among them are the copepods, which are usually no more than two millimetres long. Their importance lies in their abundance (there may be as many as 1500 per cubic metre of average sea water), for copepods form the major component of the animal plankton that helps to feed the world's fish.

One of the most fascinating things about these tiny crustaceans is their advanced method of reproduction. When mature, the male cope-pod produces a plastic substance that hardens into a bottle-like structure with a long neck. He puts the sperm into the bottle and, carrying it in his hind pair of thoracic legs which are modified into a pair of delicate pincers, attaches it to his chosen mate. The sperm then passes out of the bottle and into the body of the female where fertilisation takes place.

Mussels large and dark, and barnacles small and white, exposed at low tide. The line below which barnacles do not grow can be clearly seen.

Typically the fertilised eggs of all crustaceans hatch into a free-swimming larva called a nauplius, a minute egg-shaped creature with three well developed pairs of legs. In some the nauplius is released from the egg to swim with the plankton; in others the egg hatches at a later stage at which time the larva looks like a miniature of its parent. They are often adorned with long spine-like processes to aid floatation, but these are lost when the larvae change into adult form and sink to the bottom to take up their more permanent residence. At first, it may be hard to see why free-ranging animals like crabs and lobsters need a dispersal phase in their life cycle. The fact is, however, that the home territory of a crab is minute compared to the ocean habitat and, as for all living things, too many of one sort in one area can lead to trouble.

The barnacles are one group to which dispersion is of the utmost importance because in their adult form they are completely sedentary, fixed to their appointed spot for life. Although super-

The tide is out so the acorn barnacles have shut up shop.

opposite
The tide is in, and the acorn barnacles are feeding.

ficially they look like molluscs (and are considered as such by most people who visit the seaside), barnacles are indeed crustaceans. The problem is that most of us only see them when the tide is out and they are shut down out of desiccation's way. Once resubmerged the 'mollusc' does not begin to walk about: its 'feet' appear not at the base but at the apex of the shell and start to grasp at the water, catching particles of food and carrying them into the mouth. The 'shell' is in fact a highly modified exoskeleton, and inside it the very atypical crustacean lies on its back waving its six pairs of thoracic limbs in the air, for these are modified into food catching fan-like organs called cirri.

Barnacles are hermaphrodite, that is, each individual has both male and female organs. Nevertheless cross-fertilisation between two adjacent individuals is the order of the day, and as fertilisation is internal this means that the distance between neighbours must be less than the length of the male organ of reproduction, the penis. On the sea shore this is usually no problem because they are often packed together very closely on the intertidal rocks. There are, however, some that live a less crowded existence so they have the problem of finding another individual within mating distance. In these the male barnacle, often very reduced and sometimes even lacking a gut, gets round the difficulty by living inside the 'shell' of the much larger female.

The nauplius larva of the barnacle is released into the water, where it undergoes a series of moults, getting larger each time until it turns into another larval form, the cypris. This looks like a minute mollusc with two shells that may be held shut by a strong adductor muscle. It is the cypris that finally seeks out a new bare piece of rock on which to attach, and the antennule, supplied with a special cement gland, senses out the right spot and then fixes the animal down for good.

It is perhaps not a great jump to go from a male animal living as a commensal within the carapace of its mate, to a truly parasitic barnacle. *Sacculina* is just that, a barnacle adapted in the most fiendish way to parasitise crabs. *Sacculina*

Goose barnacles feeding. Note the long neck of the one on the right.

opposite
Look at my legs! A banded coral shrimp.

begins its life as a free swimming nauplius. Although it lacks both a mouth and gut it changes into the cypris stage, eventually becoming attached to a crab by means of its antennule. The unfortunate crab is then doomed to complete takeover by the parasite.

First the cypris rids itself of those parts of the body that are superfluous to its new parasitic mode of life, and that includes its complete trunk, appendages and all; they are simply cast off with the old cuticle at the next moult. The new cuticle consists of nothing more than a hollow dart that is stuck into the host's body and through which a mass of undifferentiated cells

Lepeoptheirus normanni, a parasite of the sunfish.

opposite
A decapod crustacean eating
Cyclothone braueri, a fish.

A trilobite (extinct arthropod) fossil.

The king or horseshoe crab.

hulls below the water line. In recent years, as ships have been able to go faster and faster, fouling by barnacles has become less for the simple reason that at speeds in excess of eleven knots they can no longer hang on. Their place has been taken, at least on the faster vessels, by seaweeds and it is an amazing fact that quite a short growth of green or brown seaweed covering the hull adds to the drag of the ship to such an extent that in no time at all it can double the fuel bill. So acute is this problem that much research effort is being directed to finding a way of keeping hulls free from fouling growth.

Barnacles are however still a hazard, not only to smaller, slower ships but especially to the inside of the large pipes that take sea water into coastal power stations for cooling purposes. The full extent of the difficulty can be realised when the immense number of larvae found in the plankton of oceanic water is taken into consideration. Out of the average of 4500 animals found in one cubic metre of water, no less than 2000 could be the larvae of barnacles. With a medium sized power station using many millions of litres of water per day . . . well, it does not take too much imagination to see that this all adds up to one big problem.

surrounded by ectoderm—all that now remains of the larva—passes into the host. It is then carried through the blood stream to become attached to the underside of the intestine. From here on, the parasite just grows and grows, sending rootlet-like extensions of itself throughout the whole body of the crab, from the telson to the tip of the great claws. All this time the main part of the barnacle grows bigger and in so doing presses onto the underside of the host's abdomen, to the extent that at the crab's next moult no exoskeleton is formed at that particular spot. The result is that the parasite protrudes out onto the undersurface of the crab, which has become no more than a living bulk permeated by the now adult parasitic barnacle.

It all sounds rather like some far fetched figment of a scientist's imagination. But not so—it is for real, and *Sacculina* is by no means a rare beast. Here is a good example of one of the main principles of evolution. Wherever there is potential, evolution will find a way to use it. The mechanisms are as bizarre as the forms of life and none more so than the parasitic crustaceans.

At one time barnacles posed an enormous problem to all ocean-going shipping for they found a very convenient home stuck onto the

Case of women's lib
sea spiders

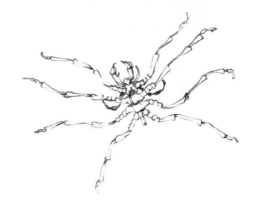

The crabs and their kin will long continue to exercise their many pairs of legs in the marine environment. In contrast, the insects with three pairs of legs and the spiders with four reserve theirs for the dry land and sweet waters of the world. There is however one group that must not be forgotten. It lies somewhere between the crustaceans and the spiders, finds its own particular leg room in the sea, and goes by the various names of the *Pycnogonida*, sea spiders, or, most descriptive of all, the nobody crabs.

Nobody crabs can be found by anybody with the patience to look. They slowly crawl about on the surface of seaweeds and invertebrates, and look like a spider with a very bad dose of rust in its joints. Their affinity with the true spiders is very tenuous and, although many of them do have four pairs of legs, more recent finds in the Antarctic have revealed sea spiders with five and even six pairs.

The nobody crabs have no body: the part in the middle of the bunch of legs consists of a

What, no body? *Pycnogonum,* a sea spider. Note the long chelophores.

A little more body — a sea spider from King George
Island, Antarctica.

Pycnogonum littorale, a sea spider.

much reduced cephalothorax (head and chest
fused together) from which the legs sprout, and
its abdomen is reduced to an unsegmented tail-
like projection. In fact the body is so reduced
that there is insufficient room for all the essential
organs—they literally overflow out into the legs
and it is not unusual to find a female with eggs
developing inside the top section of her legs.
The tail-like projection is balanced at the other
end by a proboscis that in comparison may be
very long, and terminates in the mouth. This
stiff proboscis is used to suck the juice out of
their chosen diet of hydroids and other succulent
invertebrates.

Immediately behind the proboscis three more
specialised appendages are found. The chelophores terminate in a pair of pincers, some of
which are large and present quite formidable
weapons. Next there are the pedipalps that are
often very fine and perform the delicate task of

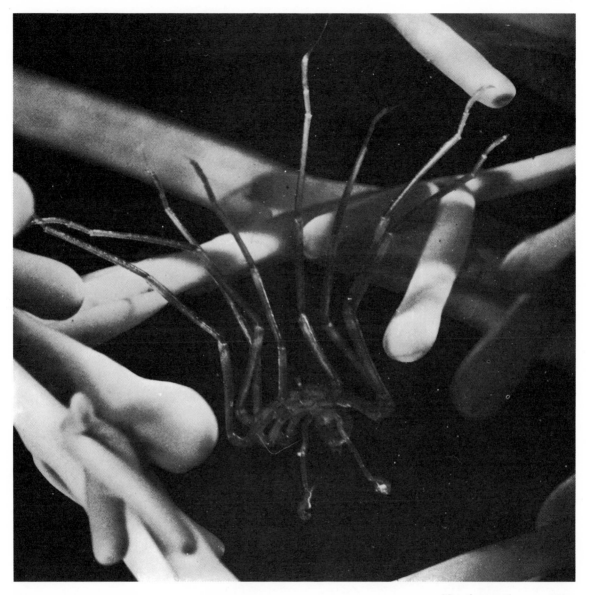

Nymphon gracile, a sea spider.

feeling and sensing the environment. Behind these are the ovigerous legs that look not unlike a smaller version of the true walking legs, and it is here we find the most peculiar feature of the group—the egg legs are best developed in the male of the species.

Mrs nobody crab lays her eggs in spherical or cake-shaped masses; her partner then collects them up and carries them about with him, hung onto his ovigers. In time he amasses such huge numbers that his movements are slowed down

even more than usual. When the eggs are developed they hatch out to produce larvae that at first glance are reminiscent of the nauplius of a shrimp. The larvae continue to hitch a ride until they eventually simply drop off to start their own free existence. Of all the members of the armoured corps of arthropods it is only the pycnogonids that have gone in for this ultra form of women's liberation.

A good foot forward
molluscs

In uptight zoological parlance the molluscs are unsegmented coelomate animals, with a head (usually well developed), a ventral muscular foot, and a dorsal visceral hump with soft skin, the part covering this hump (the mantle) often secreting a largely calcareous shell and produced into a free flap or flaps to partially enclose a mantle cavity into which opens the anus and mesoblastic kidneys, etc, etc, etc. The complete description goes on for another page and leaves you with the distinct feeling that snails could never be that complicated. The fact is that they are, for the *Mollusca* together with the *Crustacea* are the first groups of animals to show the full complement of organ systems that set the pattern for all higher organisms.

The molluscs have many features in common with both the true worms and the crustaceans,

A single-shelled mollusc, the painted topshell *Calliostoma zizyphinum.*

but among the invertebrate animals, that is those without backbones, they are in a class of their own when it comes to 'brain power'.

Of all the groups of animals to possess let alone to have evolved a complex brain, the molluscs would at first sight appear to be among the least likely. Neither the slow glutinous movements of the snails and limpets, nor the sedentary life of the oysters and clams, would seem to require it. However the above title not only applies to the single-shelled gastropods and the two-shelled lamellibranchs but also to the squids and octopuses, and it is in this latter group, the *Cephalopoda*, the majority of which are active hunters of the sea, that we find the real advances in thinking ability.

In the crustaceans it is the legs that show the main modification to function, while in the molluscs the single foot shows an equally staggering range of adaptations related to its owner's

The four basic forms in mollusc design.

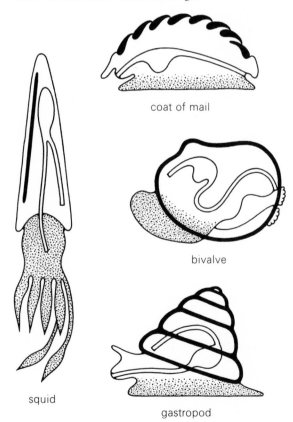

coat of mail

bivalve

squid

gastropod

way of life. Cephalopod means exactly what it says, head-footed, and there is little doubt that the squids and octopuses have the brainiest feet in the business.

Apart from this, the snails and their kin would appear to be simple animals without a trace of the segmentation shared by both the true worms and the crustaceans. Comparison of the larvae of the worms and the molluscs, however, shows very close similarities pointing at least to the fact that both groups arose from the same ancestral stock. The question is therefore raised concerning the ancestry of the molluscs: were they segmented or were they not?

One group of undoubted molluscs, the chitons or coat-of-mail shells, points firmly to the former. The chitons have an elongate foot and live much like the common limpets, browsing the intertidal seaweed gardens. However their shells are made of eight, strong, overlapping plates which, together with their mass of stiff spines produce the illusion of a large wood louse fringed with short stiff hair. The illusion is completed when a chiton is detached from the rock for it immediately attempts to roll up into an untidy ball. However, owing to the stiffness of the plates, the ball is never completed so it is possible to look at the underside of this model beast. Five pairs of gills arranged on either side of the foot again indicate a segmented animal. But that is as far as it goes for there is no trace of internal segmentation, and in all other respects chitons are advanced molluscs.

In 1952 the discourses on the ancestral mollusc were revitalised by the description of *Neopilina galathea*, ten specimens of which were dredged up by an oceanographic research vessel, the good ship *Galathea*, from a depth of 3600 metres off the west coast of Mexico. *Neopilina*, with its single shell, is even more like a limpet than the chitons are. Underneath, the small foot is surrounded by five pairs of gills and paired sets of excretory organs (very like the nephridia of the true worms), paired muscles, and paired sets of reproductive organs—definite proof of both external and internal segmentation.

No wonder the paper describing this denizen of the deep caused such a stir in zoological

circles. Here was clear evidence that at least certain molluscs are segmented, and with *Neopilina* as a starting point the theoreticians soon got going again on speculations concerning the evolution of this fascinating group of animals.

The contents of the gut of *Neopilina* showed that it enjoys a somewhat exclusive diet of radiolarians, that is, protista, served up in their own skeletons which are made of purest silica.

Apart from their spicules, radiolarians should present no problem of digestion, but not all molluscs are as fortunate with their diets. Within the group there is everything from gourmet meat eaters, through absolute vegetarians, to *Teredo*, the shipworm, an animal that feeds on ships or at least used to when they were made of wood. To date the new iron and fibreglass hulls have proved too indigestible even for the enterprising molluscs.

opposite
A two-valved shell of the spiny cockle *Acanthocardia echinata*.

A coat-of-mail shell browsing among the seaweed gardens.

Assorted radiolarians (magnification × 38).

Getting to the guts of the matter, the molluscs do show a vast range of structure and complexity in their digestive systems. The basic design is a mouth equipped with a radula covered with strong curved teeth. This is worked rather in the manner of a reciprocating chain saw, rasping away at the food and reducing it to manageable proportions. A muscular crop further grinds and mixes the food before it passes into the stomach and liver. Digestive juices are supplied from salivary glands, the cells of the stomach wall, and those of the liver, and digestion is completed within the cells lining the latter two organs. Undigested material passes to the anus via a long intestine.

The shipworm is almost unique among the animals in that its liver produces an enzyme that can attack and break down cellulose, the substance of which the cell walls of plants are made. *Teredo* can therefore make use not only of the goodies within the cells but of the actual energy stored in the cell walls themselves, hence its ability to live on a diet of dead wood. Apart from a very few molluscs, all other herbivorous animals rely on certain bacteria living in their gut to crack the cellulose for them.

The method *Teredo* uses to bore into its chosen piece of wood is to rock the small valves of the shell back and forth by rhythmical contractions of the adductor muscle fibres and so scrape the wood away. The mantle lines the burrow with calcium as it goes and at the same time provides the mouth of the tube with two calcareous trap doors that close the entrance when it withdraws its long siphons. It is a lengthy process, but as the shipworm eats the wood as it works, it is not really in any hurry.

The real malleability of the molluscan gut is shown by the diminutive *Calma* which lives out its life in the yolk of fish eggs. Surrounded by high grade food it has no need for a digestive system and lacks both crop and anus, its faeces accumulating in the liver throughout its short life.

However, in many of the more normal molluscs the massive development of the digestive system, and especially of the liver, brought about enlargement of the visceral hump. If this had not been compensated for, the mollusc in question would have become top heavy. Unequal growth of the visceral hump led to

twisting and hence to stabilisation of this ungainly part of the body, and that is how the snails got their coiled shells.

The soft skin covering the visceral hump is called the mantle and this secretes the beautiful shell. Within the shell the mantle cavity accommodates not only the gills but also the anus and the exit of the kidney ducts; and thereby hangs a complicated but fascinating tale of how molluscan evolution got in a twist while sorting out the sanitary problems of the mantle cavity.

As the developing ancestral line lost all traces of segmentation, so too did it lose the all-round mantle cavity. At first the head and foot of the gastropod molluscs occupied the front end of the shell while the mantle cavity was at the rear. With this arrangement any currents set up by

The prototype chain saw, part of the radula of a limpet.

Patella vulgata was here! The natural graffiti of the common limpet's radula track as it rasps green algae from the rock surface.

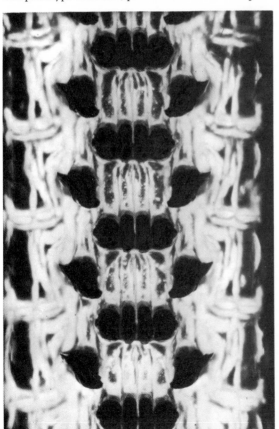

the forward movement of the mollusc would counteract any ciliary currents carrying water in to keep the gills clean and well supplied with oxygen-rich water. The simple expedient was to turn the visceral hump through 180 degrees thus bringing the mantle cavity to the fore where the currents could act together, cleansing the gill chamber. The unfortunate feature of this new twist to life was that the waste was now voided over the snail's head. However distasteful this may seem to us, evolutionarily speaking it must have been the answer, solving more problems than it created, for all the molluscs with coiled shells exhibit this 180 degree twist or torsion.

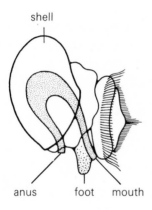

shell

anus   foot   mouth

Veliger larva of a mollusc before torsion (above) and after torsion (below).

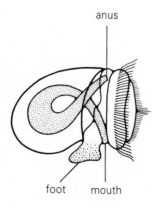

anus

foot   mouth

In other forms certain modifications are found that help to avoid the contamination of the respiratory current by waste. In both the keyhole limpet and the ormer, the shell is supplied with 'exit' holes allowing a one-way current that brings water in over the gills and out over the excretory pores. Although this overcomes at least some of the sanitation difficulties, it increases the problem of water loss, so much so that both these animals only thrive below low water mark.

One close relative of these holey molluscs does live and thrive between the tides—the very common limpet has alleviated the problems by loosing its gills altogether, their function having been taken over by the mantle itself. The respiratory current passes in over the mantle flaps and out around the foot, but of course this can only happen when the limpet is not clamped down on the rock. Life between the tides necessitates at least two periods of clamp down each day and during stormy weather it must become a full-time occupation.

The common limpet starts its life in a coiled shell. This is soon replaced by the smooth conical one, which offers much less resistance to the rush of water. Also, immensely strong muscles allow the foot to clamp down the shell onto the rock surface, safe both from dislodgement and desiccation. This clamp-down is so effective that it is often possible to see the chosen resting place of each adult as a limpet-shaped depression in the rock. From this home base the limpet sallies forth to graze the seaweed from the rocky pastures, ready at the first sign of the returning tide to rush at limpet speed, back to fix itself down on its home base.

Not all the snail-like molluscs are herbivorous; some like the whelks are voracious carnivores. Under that bulky shell is a deadly killer: its secret weapon is a long extensible proboscis fitted with a highly efficient radula that can cut its way through the carapace of a large crustacean. Once inside it just carries on rasping away at the soft flesh. In such carnivores the bulk of the digestion takes place in the stomach by means of protein-digesting enzymes secreted by

The abalone or ormer *Haliotis tuberculata* (above), and
the keyhole limpet *Diodora apertura* (right).

the stomach wall. Murex is a close relative of the
whelk, and careful investigation of its digestive
system has shown that the salivary glands and
the liver produce the full complement of protein-
digesting enzymes, exactly like those found in
more complex organisms like you or me. The
main difference is that in murex there is no
division of labour between the digestive organs;
they all produce the complete cocktail of
enzymes. Specific organs producing specific
enzymes only came much later in evolution.

It is an interesting point to remember that
without the correct enzymes, much of the food
material would remain undigested, only to pass

out with the faeces. Biochemical evolution had to precede organic evolution. Many of the lower organisms such as bacteria and fungi have the capability of utilising a whole range of the life chemicals as food, including cellulose, and this shows that the enzyme systems were evolved at a very early stage. It would appear that as the multicellular organisms increased in complexity, it became possible for evolution to build in the full biochemical armoury, opening up the potential of more and more sources of food to the evolving animal kingdom.

Other groups of gastropods overcame the problems relating to the mantle cavity by reduction and even complete loss of the shell,

thus allowing the gills to open on the outside. The fact that in all these forms the body undergoes torsion during development but untwists before reaching adulthood lends weight to the above theory of twist. This group includes the exotic sea butterflies in which the foot is carried back as two transparent flaps that are used to propel the pelagic animal through the water. Most beautiful of all are the nudibranchs or sea slugs, which have lost all vestiges of the true gills; their function has been taken over by naked accessory gill-like structures. Some of these slugs are the most bizarre looking creatures, their backs being covered with part coloured, part transparent sacs called cerata each of which

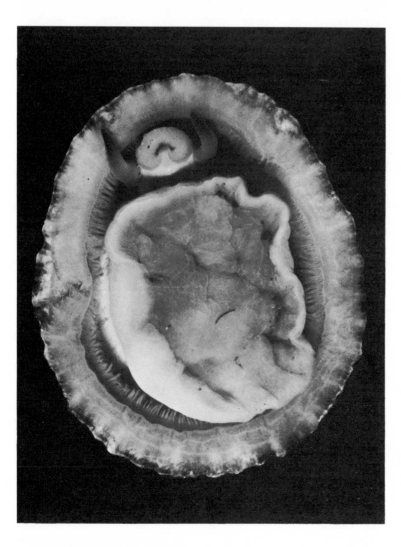

The underside of a common limpet showing the foot, mantle, and gills.

opposite
A place to call its own: the common limpet at its home base. The conspicuous scar was made by a limpet repeatedly clamping onto the soft rock.

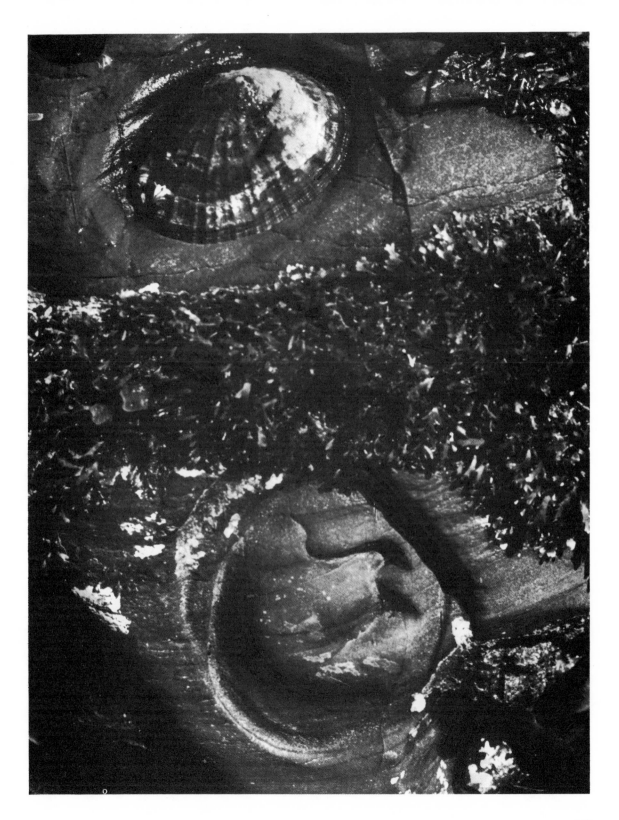

contains a branch of the 'liver'. The bright colours and striking patterns that make the sea slugs objects of great beauty, act as a warning to any would-be predators, for the cerata are full of stinging cells annexed from the *Cnidaria* on which the slugs actively feed, and each one is ready for action.

Another great group of the molluscs are the lamellibranchs. They include the bivalves, all those with shells composed of two valves that can be held clamped shut by phenomenally strong adductor muscles. The key property of the adductor muscles is the ability to maintain tone, that is, exist in a state of contraction holding the two halves together. This is quite unlike the

muscles so typical of higher animals which, as most of us know, being adapted for the performance of rapid work, are incapable of remaining in the contracted state for long periods.

The scallops are one of the few bivalves that possess both large shells and the ability to swim. Swimming is achieved by the rhythmic flapping of the shell valves, and the scallop can move in either direction by aiming the water stream with the edge of its mantle. The big adductor muscle in the scallop consists of two distinct

opposite
The sea slug *Glossidoris,* a no-shelled mollusc. If the slugs in my garden looked like this I wouldn't mind them eating the flowers.

Buccinum undatum, a deadly killer: a whelk with siphon extended.

parts: the larger is made up of the non-striped muscles typical of the lower animal groups, and this has the clamping power; the smaller portion is composed of striped muscle and it provides the swimming power. At the edge of the mantle many eyes peep out through slightly opened valves before the scallop settles down to feed in peace.

The large, flat, strong foot of the univalves was made for walking, and progression is accomplished by waves of contraction flowing across the broad, flat sole. The best way to see this is to watch a whelk moving across the side of an aquarium. Unfortunately the action of the bivalve's foot is much more difficult to observe for it is used for burrowing.

King among the burrowers, especially when it comes to speed, are the razor shells. It is almost unbelievable to approach one underwater and to see it heave its seemingly ungainly body upright and then simply disappear into the sand. Although the long shell may look bulky it is in fact quite light and fragile, as many shell collectors will have found to their great disappointment. Once upright the blood supply to the

Sea hares courting.

foot is cut off by closure of a special sphincter valve and at the same time the foot is extended into the sand. The valve is then reopened, blood is pumped into the slim-line pointed foot causing it to swell and anchor, and the muscles of the foot are contracted thus pulling the shell down. The whole operation is repeated in rapid succession so that the razor shell simply melts into the sand.

opposite
Mastermind mollusc: a common octopus stalking its prey.

A nudibranch.

The majority of the tube-like siphons that connect the buried animals to the surface are used for both feeding and respiration. Water is pumped down one and up out of the other, for unlike the wood eating shipworm they are suspension feeders. So too are the very common mussels which lie on the bottom, fixed down for most of their lives by strong byssus threads produced by the foot, and the much less common oysters, which spend their whole lives cemented to the sea floor.

In the main, the bivalve molluscs are modified to strain their food from the water stream that also supplies their bodies with oxygen. The function of respiration is taken over by the mantle, while the much enlarged gills act as a mechanism for sorting and capturing food. On entering the mantle cavity the speed of the water is checked so that the heavier particles of sand and silt fall, to be immediately carried back and thrown out through the exit siphon. The main water stream carries the lighter food particles across the great gill where a complex of ciliated tracts sorts and envelops the food in mucus, eventually carrying it to the mouth. For a long

time the bivalve molluscs were said to be filter feeders, but this is not strictly a filtration mechanism. It is therefore better to use the more modern term suspension feeders, which simply means feeding on particles held in suspension. Not all the bivalves have such large, complex gills but the majority feed in this way.

The rarity of natural pearls is a measure of how efficient this sorting mechanism is, for pearls are usually formed only when a parasite attempts to enter the mantle. The mantle soon reacts by surrounding the intruder in a small sac, which then secretes layer after layer of the pearly substance around the foreign body. The formation of pearls can be induced by the seeding of the mantle of a whole range of bivalves with foreign material.

Of all the work carried out at the Stazione Zoologica in Naples, that of John Zachary Young and his co-workers ranks among the most enlightening in that it has given us an insight into the workings of our own memories. Yet the subject of their detailed and imaginative studies was a 'lowly' mollusc, the octopus, one of the cephalopods. The reason that the Young

opposite
Is the coast clear? A scallop opens up to take a look.

A razor shell with foot extended about to do its disappearing act.

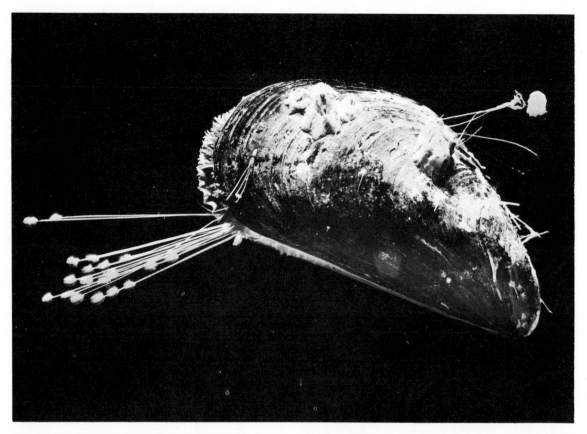

The common mussel *Mytilus edulis* attached by byssus threads to the glass of an aquarium.

team worked in Naples was not simply because it is a pleasant place in which to spend a long working vacation, but because octopuses are plentiful in the bay, and Anton Dohrn, the founder, had provided the necessary salt water workshop. Like the animals they study, the zoologists have to exploit the potential of the habitat and all it has to offer.

The work at Naples centred around the training of octopuses by reward and punishment to behave in specific ways. The classic experiments carried out showed that an octopus soon learned not to take a crab that was offered together with a specifically shaped card, for it associated crab plus card with an electric shock. In this way it was shown that they possess a memory that can last as long as two to three weeks. Further experiments showed that the brain of the octopus has some remarkable similarities, at least memory-wise, to that of us humans—'brain power' indeed.

Watching an octopus underwater with its large camera-like eyes that are even more human, both in appearance and working, than its brain, it is difficult to believe that this is a mollusc. Perhaps the large hump of its body and strange creeping movements give the game away, but absolute proof is only obtained by inspection of its internal structure.

Although a shell is completely lacking in the octopus, there is a large mantle cavity inside which are two feather-like gills. Water drawn in by the regular pumping of the mantle passes over the gills and then over the anus and out through a special muscular funnel. The only function of the gills is respiration, for a highly efficient respiratory system is needed by such a large, active, complex animal. Octopuses and squids are predatory carnivores that chase and capture their prey by means of their long tentacles, each armed with highly efficient, horny suckers. The mouth of the octopus is supplied with a strong beak that can tear at the prey if it is too large to be swallowed whole—no wonder they are the ultimate dread of all divers!

Many of the stories of sea monsters have little foundation other than a surfeit of rum. However there are some authenticated records of whales having been caught with the marks of very large squid-like suckers on their bodies, and dead bodies of large squids are found floating at sea or washed up on the beach, especially along the east coast of Canada. The marine station at St John's, Newfoundland, has latched onto this fact to become the world centre for the study of these great beasts, some of which may be as long as 30 metres.

The squids still retain their shell although it remains hidden from view inside the animal. Most people who walk the seashores will have never been lucky enough to find a stranded squid, large or small, yet many will have found the strange cuttle bones washed up with the tide. Like the giant cuttle fish, the majority of the squids live a free existence, swimming through the water and often at a great depth, where their remarkably streamlined bodies are picked out by rows of light-producing organs. It is of interest that some of them are equally as complex as those of the fish, the other group of large animals that exploit the limited potential of the black inner space of our planet.

The suckers on the feeding arm of the squid *Loligo forbesi* showing the hooked marginal rings for gripping the prey.

Forward propulsion is afforded by undulation of the fin-like extension of the mantle which drives the squid through the water, tentacles first. A group of squids swimming together looks not unlike a well trained flight of animated paper darts as they bank and turn in complete harmony. However, if danger approaches more rapid movement is produced by jet propulsion, water being shot out from the funnel by violent contractions of the mantle. As a further aid to the disappearing act, ink is secreted into the jet stream thus laying an effective 'smoke screen'. The funnel can be pointed in any direction so the animal may move forwards or backwards.

The most striking adaptation to life as an active predator is the squid's ability to change colour, and so rapid are these changes that waves of different colour can actually be seen to pass over the surface of the body. This rapid camouflage must be equally useful both in stalking prey and in defence.

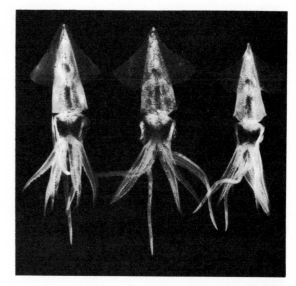

A squid formation team.

A cuttle fish swimming slowly by gentle undulation of its mantle flaps.

The delicate shell of the paper nautilus *Argonauta argo*,
an octopus with an external shell.

These active predatory molluscs have several features in common with the fish to which they are not related (fish being vertebrates, animals that have backbones). The eye of the octopus and squid is very similar to that of the vertebrates, both in structure and in function, but as each was developed in an entirely different way they cannot be regarded as the same, simply as cases of convergent evolution. In addition, the squid's system of internal cartilaginous supports, some of which enclose and protect the brain, is very like the brain box of the vertebrates. Both these facts are good examples of two unrelated groups producing similar structures for similar functions by entirely different evolutionary routes.

An exciting find for any beachcomber on tropical shores is the large coiled shell of the nautilus, another cephalopod. The shell is at once recognisable by its red-brown pattern and the way the end opens into a shallow cavity, looking not unlike a mother-of-pearl hand-basin complete with central drain hole. Unfortunately, the true beauty of the shell is hidden within, and to see it properly it must be sawn in half straight down the 'drain hole'. The shell consists of successive chambers, each one smaller than the previous and each penetrated by a tube or siphuncle that leads from the 'drain hole' right through to the smallest chamber set in the middle of a perfect spiral (see pages 56–7).

When living, the mollusc, which looks not unlike an octopus, sits inside the largest chamber, although an extension of the visceral hump passes right through the siphuncle to the centre.

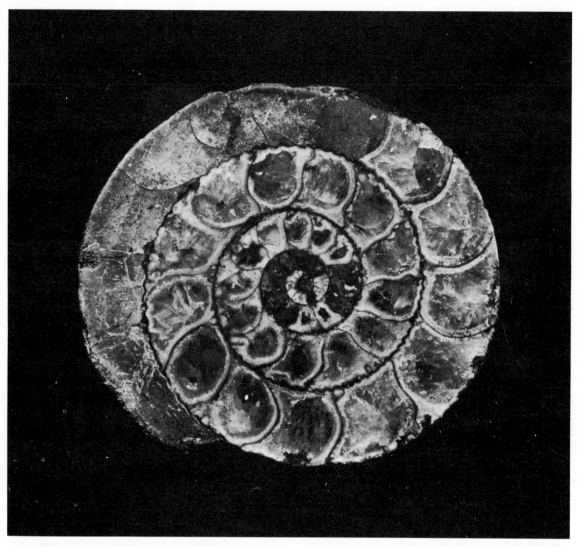

An ammonite fossil sectioned and polished.

The function of this living connection is probably to help keep the smaller unused chambers full of air thus buoying up the shell in the water. Like his cousins, the nautilus is free swimming and lives the active life of a predator. Apart from the external shell the other primitive feature possessed by the nautilus are four gills and four kidneys, just giving a hint perhaps of segmentation.

The nautilus is the last in a very long line of cephalopods that have roamed the seas of the world for more than 500 million years, being dominant forms for more than 100 million years. Their large chalky shells have ensured them a detailed fossil record, and some of the ammonites found are gigantic, measuring more than one metre across. One presumes that they lived a life not unlike that of a nautilus today, producing chambers into which they moved their typical head-foot as they grew successively larger and larger. It would be interesting to know whether their brains matched their undoubted brawn.

The problems of standing still

sea urchins, sea stars, etc

At some stage in their life cycle all animals exist as a single cell, as if they were re-living one of the earliest stages in evolution when life was represented by nothing more complex than the aptly named protozoa or first animals. In the same way that the single cell of the protozoan must contain all the information necessary for its life, growth, and reproductive processes, so must the fertilised egg of, for example, the blue whale contain all the information necessary for the life, growth, and reproduction of this the largest animal that has ever lived.

The early stages of division of the fertilised eggs of all animals follow a remarkably uniform pattern: the first division produces two cells, the second four, and so on until a ball of cells, the blastula, is formed. Up to this stage most animals look alike, for the genetic information is still hidden from even the expert's view. From the blastula stage onwards, many changes take place making it possible for the embryologist to tell which type of animal he is looking at, for the hidden information is translated into the various forms typical of the development of specific animal groups and finally into the specific animals with their own unique characteristics.

If a blastula is cut straight through its centre in a number of planes, two mirror image halves will be produced each time. The blastula therefore exhibits radial symmetry, thus giving it just about the ideal shape from which to begin making anything. In most animal groups further development goes hand in hand with the process of modification, producing an elongate structure with an upper (dorsal) and a lower (ventral) side that differ markedly from each other. The resultant living form can only be cut in the one plane to produce two similar halves and is therefore said to be bilaterally symmetrical. Human beings have a basic bilateral symmetry, with an upper and a lower side, although as we stand upright we usually tend to think of them as back and front. Although it is not suggested as an empirical experiment, it would be quite possible to divide yourself into two halves, right and left, that are exact mirror images except for the displacement of certain internal organs.

Development of this basic two-sidedness has made all sorts of things possible, especially control during rapid locomotion. Flatworms, true worms, snails, insects, crustaceans, and all the vertebrate animals are bilaterally symmetrical. There is no getting away from the fact that once evolution made the change it became the key to success. However, one group of animals appear never to have made the radical change; they stuck with the more primitive radial symmetry of the blastula and it could be said that in their own way they have 'had a ball' of success.

This group is the *Echinodermata* or spiny skinned animals, and as most of them possess an outer skeleton of chalky plates they have left their long and detailed story in the fossil record. In fact the oldest group of the echinoderms was for a long time known only from fossil material, but ninety years ago a dredge brought up a live specimen from the abyss. It is not hard to imagine the excitement on the ship when the first

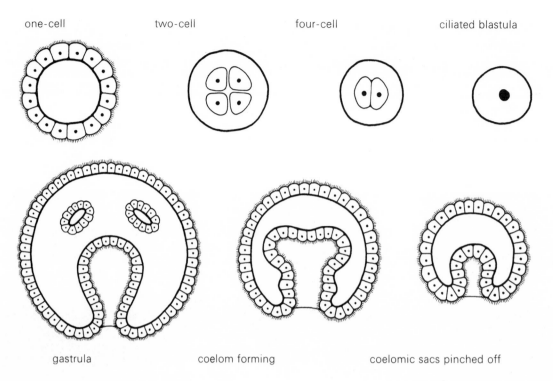

one-cell   two-cell   four-cell   ciliated blastula

gastrula   coelom forming   coelomic sacs pinched off

Single cell, through ciliated blastula, to coelom.

living sea lily, or crinoid, came up to the light of day, a member of a group that had supposedly been extinct for many millions of years.

However, more of the sea lilies later, for the easiest way to understand the group, and especially the limitations imposed by, and the potential of, radial symmetry, is to look at the contemporary echinoderms and in particular at the common sea urchin. The sea urchins do not hide their symmetry; indeed why should they? They have been exhibiting their radial success story in the sea for hundreds of millions of years, and they are still going strong, almost unchanged. The ways in which they have achieved this are numerous but several stand out as being among the most important.

First is the outer armoured skeleton which is well supplied with highly mobile and highly efficient spines. The swimmer who has had the experience of standing, or worse still sitting, on one will know this only too well. Second, although they are basically browsers, they are also omnivores, so that when pushed they will eat anything. Third are the marvellous myriad of tube feet. And last but by no means least, the urchin moves about only very slowly. If there is one thing that all us higher, more advanced organisms could learn from the echinoderms, it is that rushing around never did anyone any good, at least not in the long 'run'.

The urchin goes about its business very slowly indeed, carried forward by the tube feet all working in unison and helped along, especially among the rocks and the seaweeds, by the long mobile spines. However, there is one serious problem that all slow moving animals must face: the slower you move the more difficult it becomes to get away from would-be hangers on. This danger is especially acute for animals like the sea urchins and starfish that breathe through much of their outer surface, for too many epiphytes (plants and animals hitching a ride) would result in suffocation. Apart from numerous skin gills, the surface of the urchin is well supplied with small, extremely efficient pincer-like organs, the pedicellariae, whose function is to deal with easy riders. Anything landing on the urchin is immediately grabbed and removed by the batteries of tiny pincers that together make the finest bug rake in the business.

One of the commoner sea urchins in European waters is *Echinus esculentus*, the second name meaning exactly what it says—edible. Its common name is the sea egg. However, to crack open the shell and feast on the delights of a radially symmetrical meal would be rather a disappointment because there is little inside the shell, and what there is tastes not unlike dilute sea water. The edible part consists of the ripe eggs and the coiled intestine that connects the mouth at the bottom of the shell with the anus

An edible sea urchin showing the long tube feet.

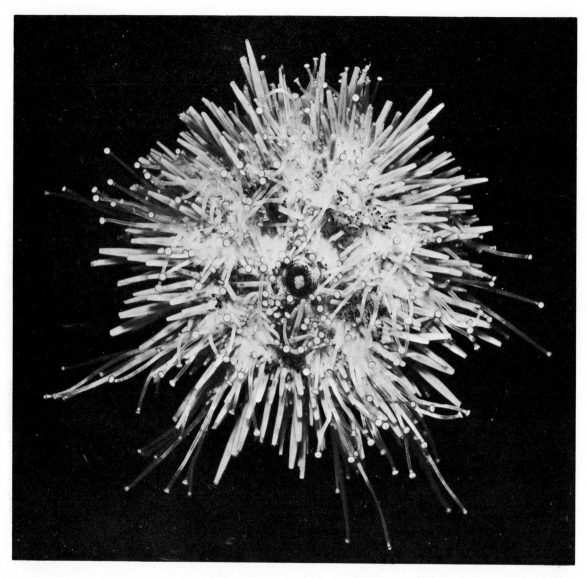

The underside of a sea urchin *Psammechinus miliaris* showing the suckers in action and the central mouth with five strong teeth.

at the top. It needs its long intestine because the vegetable matter on which it feeds requires a great deal of digestive action. Urchins are efficient browsers and an army of them moving slowly through a patch of seaweed has much the same effect as a row of combine harvesters. The famous husband and wife team of diving biologists, Joanna Kain and Norman Jones, have done some interesting experiments off the Isle of Man in the Irish Sea. They showed that removal of the sea urchins from an area below the normal range of the kelp

seaweeds allowed the kelp beds to extend into deeper water—absolute proof of their grazing efficiency.

The mechanism the urchin uses for mowing the seaweed lawn consists of five large teeth. These are worked by a complex of muscles attached to a peculiar skeletal framework that looks not unlike a Victorian bandstand and

Aristotle's lantern, the name given to the complex mechanism that works the urchin's teeth.

rejoices under the name of Aristotle's lantern. It is not unusual to find these strange structures among the flotsam of the strand line.

The exact function of the paddle-like appendages of the slate pencil urchin is hard to say. Having watched them there is no doubt that they are used in locomotion, although whether they aid or hinder is a matter for conjecture. On the other hand, the appendages of both the sea

potato and the sand dollar are very functional. The long, bristle-like spines working in unison with the long, pointed tube feet can quickly bury the animal, which literally dissolves into the sand where it lives, leaving no trace whatsoever.

From sea urchins to sea stars may seem a big jump. However if you take a close look at them side by side, it is apparent that the latter is just a flattened version of the former. The five double ranks of tube feet that radiate out over the urchin are a similar sort of structure to the five

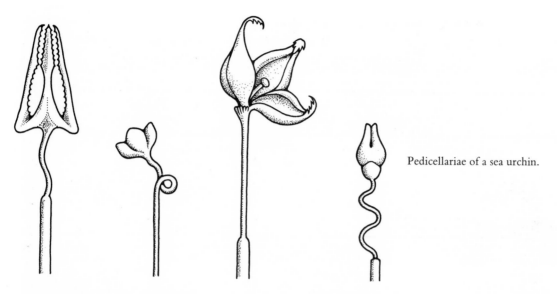

Pedicellariae of a sea urchin.

arms of the common sea star. (Most of us seem to call them starfish but as it is difficult to imagine anything looking less like a fish, sea star is really more appropriate.) The rest is also very much the same, except that the sea star is a carnivore; it therefore does not need, and indeed does not have, a long coiled gut. It does however have a most peculiar stomach that can not only

be turned inside out but in the process can be protruded out through the mouth to envelop the edible parts of its prey.

A ravening horde of sea stars can clean the mussels off a beach in no time at all. In order to understand the method by which these efficient predators feed it is necessary first to understand the workings of their wonderful tube feet. When

A slate pencil urchin. Do the 'paddles' help or hinder their movements?

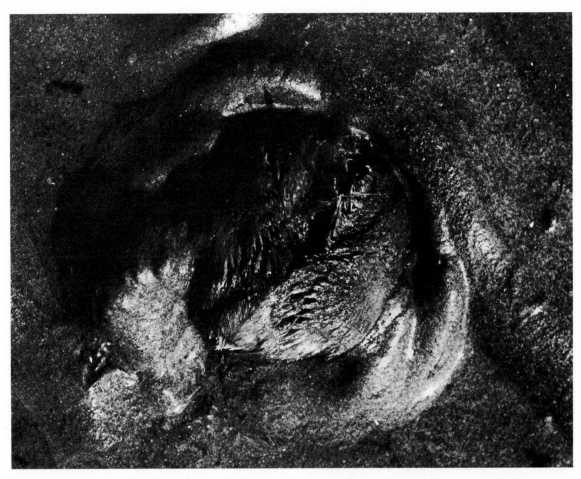

A heart urchin just about to disappear down into its sandy home.

a large sea star attacks a scallop bed, the frenzied scallops flap about in an ungainly way, gnashing at the intruder, and looking not unlike sets of flying false teeth. Once the sea star has homed in on one particular scallop and locked onto its shell, the end for this unfortunate beast is not too far away. The success of the whole operation depends on the myriad of tube feet worked by a complex hydraulic system.

The hydraulic system connects to the outside via a pore; this is covered by a sieve plate that lies almost at the centre of the sea star on its upper side. Water is drawn into the system through the sieve plate by the beating of cilia lining the stone canal, and in this way the hydraulic system is kept full. Pressure is applied by the contraction of the ampulla, which forces water into its adjacent tube foot, pushing it forward

Water vascular system of a sea star.

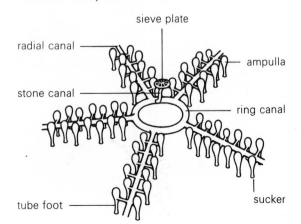

and into contact with the shell to which it becomes attached by its sucker-like end. The muscles in the foot now contract, pulling on the shell of the scallop. One of the tiny tube feet alone would not be much use against the muscles of the scallop as anyone who has tried to open one will know; hundreds of them working together is a different matter and here is the real secret of success. The scallop has only two muscles to hold its shell tight shut and although made of many separate cells they all work together so they quickly tire. The sea star uses its batteries of tube feet in relays; when one is tired, another takes over. It is an unfair struggle that the scallop always loses. Once the shell is pulled open the sea star's protrusible stomach completes the job.

All the early spiny-skinned animals were sessile, fixed to their appointed spot for life by a support of varying length. The original sea lilies were not unlike a sea urchin stuck on top of a long stalk. The contemporary ones, all of which live in the deep dark waters of the abyss, are more like sea stars on stalks. The nearest thing to them that can be found by itinerant beach combers is the segments of fossil crinoids eroded out of crinoidal limestone (to which they give their name). On certain beaches on Holy Island in the North Sea these are often quite abundant and are known by the local name of St Cuthbert's beads (St Cuthbert, who brought Christianity to the kingdom of Northumbria, once lived there). They are also found in other parts of the world, wherever crinoidal limestone outcrops.

One feature regarded as the trade mark of the echinoderms is the 'furrow' picked out by the rows of holes through which the tube feet protrude. The name given to this distinctive furrow is the ambulacrum, meaning a place to walk. It is easy to imagine the link, the two ordered rows of tube feet looking just like trees lining an avenue. The ambulacral grooves mark out the basic pentamerous symmetry found in the sea stars and it is thus possible to divide the star into mirror image halves by cutting along any of these five grooves. It is necessary to use the word 'basic' because some of the sea stars have more than five ambulacra, but they are always in multiples of five.

The sea stars have another fundamental property: if you do cut them in half, each half can regenerate a new one. In fact quite a small part of one arm can sit tight and make four new ones. This once proved a great problem to oyster fishermen, who used to rake the invading sea stars from their valuable oyster beds, chop

A fossil sea lily—a sea star on top of a chain of St Cuthbert's beads.

opposite
Close-up of the arm of a crown-of-thorns sea star to show the gripping feet.

opposite
A satisfied sea star at grips with its next meal.

A sea star regenerating lost arms.

them, up, and throw the bits back into the sea—with obvious catastrophic results. Fortunately they now know better. It has taken marine biologists much hard work to even begin unravelling the marvels of the echinoderms' regeneration, an ability that must have been a major factor in helping them exploit the potential of the seas for so long.

Echinoderms are found today in every ocean of the world and they come in five main types. The sea lilies (or crinoids) are the weirdest and most primitive of all the spiny-skinned animals. Fixed to their long stalks, their tube feet are not therefore needed for locomotion. Instead they help perform the process of respiration, their thin walls allowing the ingress and egress of oxygen and carbon dioxide. Few people, even marine biologists, ever see a living sea lily and so we know only very little about their way of life. Errant stars are the shallow water relatives of the lilies and like them they start their life on a short stalk. However very soon the 'flower' part of the lily detaches itself from the stalk and swims away by the rhythmic waving of its

tentacular arms. Sometimes they are so abundant in the shallow waters that it is difficult to see the seaweeds on which they congregate.

The commonest of all intertidal echinoderms are the sea stars or starfish. They come in all

The common sun star *Solaster papposus*.

165

shapes, sizes, and colours, from the common pink five-armed variety to the many-armed crown-of-thorns, which lives by actively grazing the polyps of the coral animals. The crown-of-thorns has recently become somewhat infamous as it is thought by some people to be destroying the coral reefs in certain parts of the world. Firm favourites of mine are the cushion stars, which live up to their names in the shallow waters of most tropical reefs. They look just like scattered hassocks, each slowly moving over the sand ready for a devout skin diver who wishes to kneel in prayer.

Serpent stars, or brittle stars, are a rather peculiar group of animals. The first name arises from the many highly mobile arms, each festooned with tube feet that make them look

Porania pulvillus, a cushion star.

Antedon bifida, a feather star.

A brittle star snaking off across the sea bed.

very much like serpents as they snake their sinuous way across the bottom, and especially when they emerge out of cracks in the rock. The second arises because their arms are brittle and break off very easily. However they can be regenerated very quickly, rather like the formidable serpent Hydra of Greek mythology. The place to see brittle stars in action is in the shallower waters of the continental shelf where they lie in writhing piles many deep, all enmeshed in some strange type of benthic condominium.

It is among the sea urchins that the process of evolution has produced the most bizarre forms. Undoubtedly the long sharp spines are for protection, whereas the short blunt ones are used for locomotion and for continually excavating holes in the living coral rock. As for all the other permutations and combinations of form, shape, and size, it is perhaps just as well to think that at least some of them were not evolved to do any specific job, but that they just happened.

Last, but by no means least, the sea cucumbers have done the most to keep up with the main stream of evolution in that they have become elongated. Hand in hand with this elongation

Cucumaria frondosa, a sea cucumber showing rows of tube feet.

The head of a sea cucumber with tentacular crown just emerging.

has gone an increase in mobility, and the longest and thinnest like the snake cucumber can move quite quickly; indeed it does look much more like a snake than an echinoderm. Increased mobility has decreased the need for pedicellariae and so in the snake cucumbers they have been replaced by tiny anchor-like hooks that must help greatly in traction but which give the surface of the animal the feel of nylon 'adhesive' clothes-fasteners. Long they may be but they have not lost their radial symmetry as the crown of tentacles around the mouth shows only too clearly.

Sea cucumbers are fascinating animals, quietly going about their business feeding on the bottom detritus like programmed vacuum cleaners. With neither spines nor armour for protection, they rely on a rather unique escape mechanism. When danger looms they simply eject part of their gut, the predator goes away satisfied with

opposite
I'm dreaming of a white cucumber: *Cucumaria saxicola* climbing onto red seaweed.

A snake-like sea cucumber.

a long, thin meal, and the cucumber sits tight and regenerates a new one. What a way to get rid of a gastric ulcer! Although they may not look very tasty, sea cucumbers are eaten by gourmets as a soup, under the name bêche-de-mer.

Perhaps just sitting tight has been the key to the long success of the rather decorative echinoderms. For, in an environment that has remained stable throughout the whole of evolutionary time, they have been doing just that and have overcome the associated problems. The continents have drifted and some oceans have even dried up, but throughout this long period there have always been both an abyss and large areas of shallow water; and there have been echinoderms exploiting the potential of these habitats for well over 500 million years—not a bad record by any standards.

A circle of animals
some peculiar animal groups

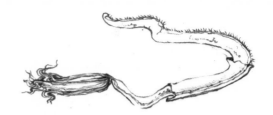

Taxonomy is the science of classifying things, the science of discovering relationships and, on the basis of those relationships, grouping things together; in short it is the science of pigeonholing. It is possible to classify anything, so why not start with me. One way in which I could be classified is as follows:

my Christian names are	David James
I am a member of the	Bellamy family
I was born in London and am	Cockney
my passport states that I am	British

British, yes, but there are at least 55 million others in that category so I am just one of a very large crowd; Cockney, well that narrows it down a little; Cockneys called Bellamy, now we are getting somewhere (although there are 103 Bellamys in the London telephone directory and of those, 8 have the initial D). But a British Cockney answering to the name of David James Bellamy—surely that must be a unique pigeonhole. However there may be, and indeed probably is, somewhere in the world another person who exactly fits that description, so to be absolutely sure I would have to be narrowed down still further by adding more specific characters, thus creating smaller pigeonholes.

So why not go the whole hog and use one, or rather ten, characters that are unique—my fingerprints. Although these are mine and no one else's, they are not much use except to a fingerprint expert. And how could that complex of patterns ever be used as a description in words?—squiggle, squiggle, U-bend, straight—absolutely hopeless! David James Bellamy, British Cockney, is much better, the only problem being that it could be ambiguous.

The taxonomist looks at 'things', decides on how and why they are related, and then erects a classification, a taxonomy giving him a language to use when discussing those 'things' and their interrelationships. It is a very exacting science and to become an expert taxonomist the essentials are to have great patience and an eye for detail.

The real problem comes, however, when something new is found, say a new animal.

One of my finger prints.

Careful and detailed study allows it to be pigeonholed as closely as possible and then finally a new, absolutely specific pigeonhole must be created for it. The animal is given its very own Latin name, a detailed description is published, and the specimen (or specimens) on which the description is based is placed in a museum where it is cared for as a type specimen. Thus the taxonomy of both plants and animals is based as far as possible on unique characters found in and described from the type specimen. This practice gives us an exact language in which we may discuss the problems of natural history and of evolution.

This chapter is basically about a very recent problem of taxonomy and about one of the most exciting zoological finds of the twentieth century, a most peculiar group of animals that were later christened the beard bearers.

At the turn of the century a Dutch ship was dredging in the deep waters off Indonesia and among the samples that came up in the dredge were about twenty-five specimens and a number of empty tubes of a most peculiar animal. The French taxonomist Maurice Caullery studied, described, and named it *Siboglinum weberi*, and that was about as far as he went. He could not decide in which larger pigeonhole it should be placed. The most puzzling thing was that the animals appeared to lack a gut, that is, they had no digestive tract. Further finds of these peculiar animals confirmed the complete absence of a gut and that they were indeed something new and very exciting. It was not until 1949 that another taxonomist took the plunge and concluded that these tube dwelling, worm-like animals were members of a hitherto unknown animal group, to which he gave the name of *Pogonophora*, meaning beard bearers. These animals were later shown to be representatives of a completely new phylum—an amazing and extremely important modern finding. A phylum is a pigeonhole of very high rank, that is, a primary division of the animal kingdom. For example, one phylum, the *Chordata*, contains not only all the vertebrate animals (fish, amphibians, reptiles, mammals, and birds) but also the sub-phylum *Tunicata*.

Having said that, unfortunately even now not all that much is known about the beard bearers themselves. The main problem is that the majority of them live only at very great depths in the abyss, where it is almost impossible to study their way of life. In addition, although they may be quite long, they are often less than

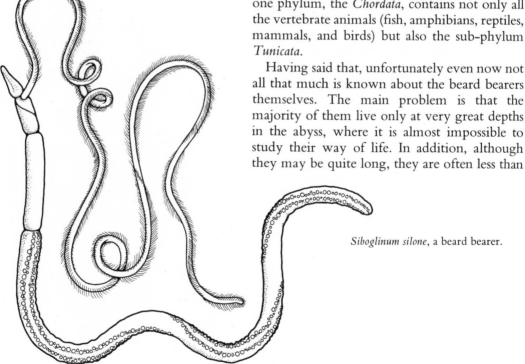

Siboglinum silone, a beard bearer.

one millimetre thick, so they present a problem, a long thin problem. One of the longest, which also has the extensive name of *Zenkevitchiana longissima*, reaches 35 centimetres but lives in a tube that can be four times that length.

The tube is partly buried in the detrital ooze that covers the floor of the ocean deeps, and the pogonophore protrudes its front end and especially its crown of beard-like tentacles from the tube for feeding. Once they have obtained the food the great mystery is how they digest it, for without a gut it would seem impossible. However it should be remembered that the 'food' filters down into the abyss from above and may already have been partly pre-digested by other animals on the way down. Furthermore there are plenty of bacteria on the floor of the ocean and once the beard bearer has collected its food, it is not inconceivable that it just retires into its tube and gets on with the job, ably helped by a host of bacteria. Recently, evidence has been obtained indicating that the process of digestion is external, the products of the enzyme action being absorbed through the tissues that make up the beard.

In an attempt to understand the interrelationships of the beard bearers and other groups of the animal kingdom, the German taxonomist Ulrich placed them in a *Tierkreis*. Literally translated this means an animal circle, and it is a circle that includes some other highly peculiar groups. The reason for putting all these groups together in such a circle is because they all lack chitin, a complex substance that forms the structural part of many invertebrate animals. Instead they all possess another chemical, creatine phosphate, in their make-up. Creatine phosphate is all-important in the chemistry of respiration of most vertebrate animals, therefore the members of the *Tierkreis* are a sort of evolutionary cocktail, a mixture of invertebrate with a dash of vertebrate characteristics.

Ulrich includes the echinoderms and protochordates in his circle; both of these groups are of great importance and are dealt with elsewhere. The circle is however closed by three groups: the phoronids, which appear to have no widely accepted common name, the bryozoans or moss animals, and the brachiopods or lamp shells. It is a very fitting trio with which to close Ulrich's circle as the members of all three groups are related by the possession of ciliated tentacles situated on a horseshoe-shaped fold encircling the mouth, which is termed the lophophore. In all three groups it is a mechanism for collecting food.

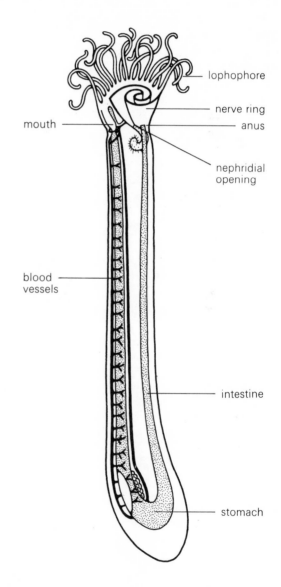

A phoronid cut in half to show the internal structure.

Two types of sea mat growing on seaweed: *Electra pilosa* encrusts the weeds, and *Scrupocellaria scruposa* is the tufted ball in the centre of the picture.

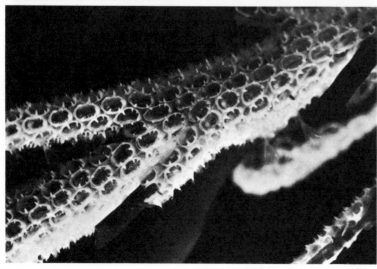

Close-up of *Electra pilosa*.

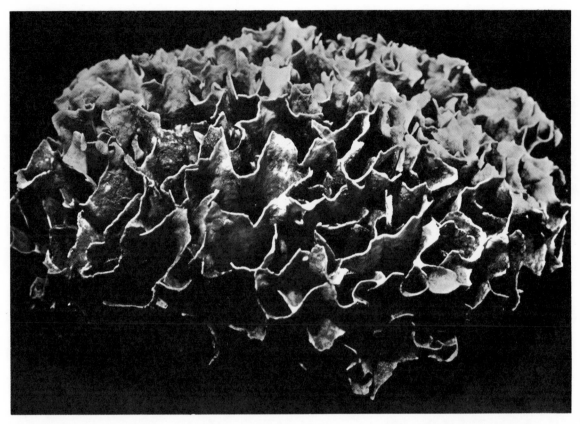

Honeycomb coral *Pentapora foliacea,* a colonial bryozoan.

The phoronids are tube-dwelling worms that look not unlike the beard bearers. However unlike them, their tentacular arms include the mouth as part of the same structure, the combination being called the lophophore. Also, they have a well defined gut, and live in shallow water. The tentacular crown collects particles from the water current created by the cilia covering the tentacles, and the food particles are trapped in mucus and carried downwards into the mouth.

Although the three groups share the same feeding mechanism, in gross structure they could not be more unlike and indeed they are very probably unrelated. As their name suggests, the moss animals look like, and for a long time were considered to be, plants. Rondelet, who was probably the first person to describe them in detail, called them *giroflée de mer* (gillyflower of the sea) way back in 1599 and thus they were

regarded until the early nineteenth century. Bryozoans live in colonies, which can be just about any shape, and all the microscopic members of the colony, and there may be many thousands, are produced by simple division from a single progenitor. Within each small cell or flask on the colony is a tiny individual that bears at its anterior end the crown of tentacles that surrounds the mouth and is used for food collection. The tentacles of course protrude out from the cell, but owing to their minute size they are very rarely seen by anyone other than the expert with his microscope.

The final group in the circle, the lamp shells, are not so difficult to see and this is especially true of some of the extinct members that lived some 500 million years ago and were up to 40 centimetres long. Like most animals with shells

opposite
Finest lace, a colonial bryozoan
Retepora cellulosa.

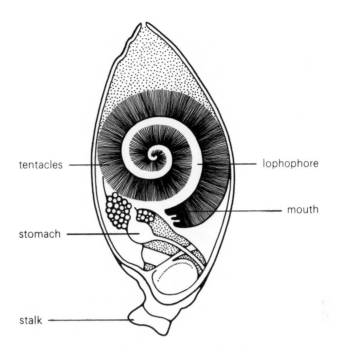

tentacles

lophophore

mouth

stomach

stalk

Lamp shell showing structure and
lophophore.

they have left a good fossil record that tells us they had their evolutionary hey-day back in the Cambrian to Devonian periods, between about 350 and 550 million years ago. Unfortunately there are no monster lamp shells in existence today. The group as a whole has been scaled down, perhaps to gain survival, but the 260 plus contemporary forms are quite widespread, being found from the sea shore down to the abyssal depths.

It is easy to see why the early taxonomists placed the lamp shells with the clams and oysters in the phylum *Mollusca*. However the lamp shell's shell is very different from the mollusc's: the two halves are in fact a top and a bottom not a left and a right as in the molluscs, and they are not hinged but are simply held together by muscles. There should really be no trouble in placing them in their correct pigeon-holes (at least the modern ones) because once the shell has been opened the tell-tale lopho-phore can be seen, with its two coiled tentacular arms that the original investigators thought the lamp shell used for locomotion. Indeed this is how they got the name brachiopod, which simply means arm-footed.

The story of the animal circle certainly tells us many things about both taxonomy and evolution. It warns the would-be taxonomist not to jump to conclusions without sufficient evidence, and it shows that once evolution got onto a good thing like the lophophore, it stuck to and made the most of it. There seems little doubt that the lophophore is an efficient mechan-ism for capturing food. Just how closely the beard of the pogonophores is related to the tentacular crown that characterises the other members of the circle is more a matter of conjecture as is the exact position of the mem-bers of Ulrich's *Tierkreis* in the evolution of the animal kingdom.

177

Striking a 'cord'
protochordates

The last chapter closed one of the great books of evolution, the book of the animals without backbones. Acellular, cellular, tissue, and organ are the four grades of construction represented by diverse groups of invertebrate animals still exhibiting their varied success stories in the modern sea, every one fitted to play its own particular role in the economy of the oceans.

Within each group of complexity and within each basic form we have seen relationships that make use of the full potential of the marine habitat: herbivores that eat plants, carnivores that eat herbivores, saprophages that feed on the material of decay, and parasites that sap the life of others—all of them able to make use of whatever potential evolution puts their way, but all of them limited in some way by the lack of a key feature. It is of interest that the organ grade of construction opened up the possibilities of life on the dry land, for each of the major groups from the flatworms onwards produced terrestrial forms, the ultimate prize going to the insects.

Early in the development of the many-celled animals, in fact at the dawn of the evolution of the third layer of cells or mesoderm from which the bulk of the complex organs of the body were later to be formed, a definite split took place in the evolutionary stock. The group destined to give rise to the crustaceans produced its third layer of cells by direct budding from the primitive gut wall, or endoderm. In the other group the mesoderm was formed as definite pouches, outpushings of the primitive gut itself. At first sight, this is not much of a difference, but it is a very basic one: in the former the coelom had to be formed *de novo* as a special split in the new masses of mesoderm, while in the latter the coelom was ready made, part of the actual cavity of the gut. Just why this distinction was so important we do not know, but it was the latter group that gave rise to the echinoderms, with all their problems of standing still, and the chordates, the huge group that includes the fish, amphibians, reptiles, birds, and mammals.

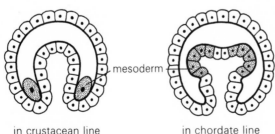

in crustacean line | in chordate line (including man)

mesoderm

The two ways that the mesoderm was formed.

opposite
One possible scheme for the arrangement of the main animal groups on the tree of evolution. Ulrich's circle of animals or *Tierkreis* (page 173) is shown as a reminder of the problems of taxonomy and the fact that there is still much work to be done.

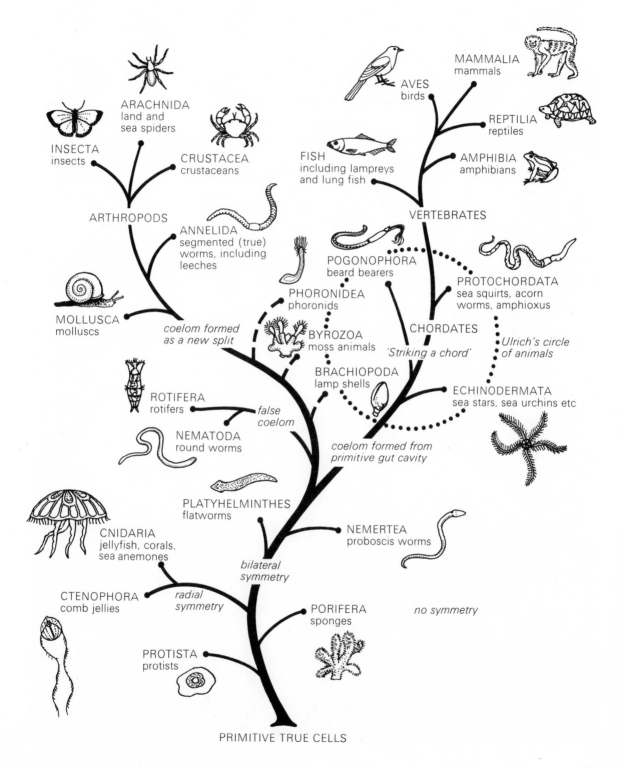

ARACHNIDA
land and
sea spiders

INSECTA
insects

CRUSTACEA
crustaceans

MAMMALIA
mammals

AVES
birds

REPTILIA
reptiles

FISH
including lampreys
and lung fish

AMPHIBIA
amphibians

ARTHROPODS

ANNELIDA
segmented (true)
worms, including
leeches

VERTEBRATES

POGONOPHORA
beard bearers

PROTOCHORDATA
sea squirts, acorn
worms, amphioxus

PHORONIDEA
phoronids

MOLLUSCA
molluscs

*coelom formed
as a new split*

BYROZOA
moss animals

CHORDATES

'Striking a chord'

*Ulrich's circle
of animals*

BRACHIOPODA
lamp shells

ROTIFERA
rotifers

*false
coelom*

ECHINODERMATA
sea stars, sea urchins etc

NEMATODA
round worms

*coelom formed from
primitive gut cavity*

PLATYHELMINTHES
flatworms

NEMERTEA
proboscis worms

CNIDARIA
jellyfish, corals,
sea anemones

*bilateral
symmetry*

CTENOPHORA
comb jellies

*radial
symmetry*

PORIFERA
sponges

no symmetry

PROTISTA
protists

PRIMITIVE TRUE CELLS

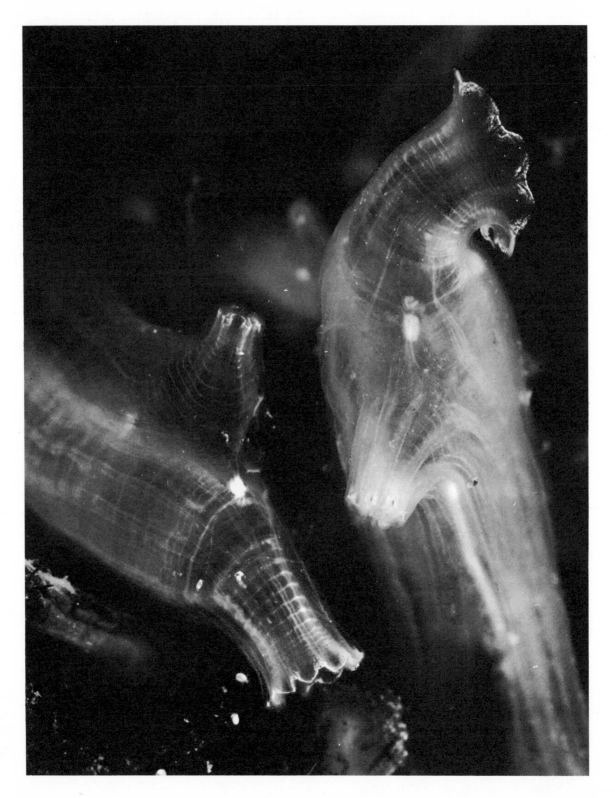

Tempting as it is to link all the various groups into a single line of increasing complexity, it would be wrong. Each group is an entity, each an experiment in evolution in its own right, a branch on a tree of diversity and not a rung in a ladder of advance.

Perhaps the most fascinating question is, why the need for complexity? What advantage has a vertebrate got when compared to a protist? A protist can perform all the functions of a living organism: it moves, respires, excretes, senses, and reacts to the environment, and it can reproduce itself. Biologically speaking, man himself can do no more than that. The only real difference is that the vertebrates do it on a much grander scale. It is almost as if once the chemistry of life had come into existence, evolution had to happen. The first forms of life created potential for others, and so it went on through evolutionary time, gathering momentum and creating more potential; more plants and animals together exploiting to a greater extent the resources of this planet earth.

The fact that the dominant life form in the sea was once the great shelled cephalopods is of little consequence; it was just a phase of development. They, like the fish today, were ultimately dependent on the photosynthetic protists of the plankton in all their diversity. The take-over by the fish in more modern times is simply due to the fact that they are at present fitter to play the same role in the ocean. Perhaps they are more efficient members of the web of life, or perhaps that web has changed in other ways since the time of the ammonites—nobody knows.

This to me is the fascination of biology. The more information we have, the deeper we can delve into the workings of the evolutionary system, and it is a sobering thought that we are probably the first products of the system to have that ability. The seed of our evolutionary success was sown at least 700 million years ago

opposite
Solitary sea squirts *Ciona intestinalis* showing their inhalant and exhalant siphons.

A family of solitary sea squirts.

opposite
Botryllus schlosseri, a colonial sea squirt. Each individual of each sub-colony has its own mouth but shares an exhalant pore with the other members of the group.

when some ancestral form budded off its mesoderm in the form of small pouches. But how do we know? where is the link?

The clue lies in the sea, but in a most unlikely animal, not only to look at but also in the fact that part of it is made of a substance very like cellulose, which is one of the main distinguishing features of the plant kingdom. The animal, or rather animals, in question are the ascidians or sea squirts.

The majority of the sea squirts seen between the tides are colonial forms, each looking like a spreading blob of stiff jelly with a repeating 'flower' pattern consisting of raised 'petals' with a central hole. In deeper water the solitary forms range from stiff transparent to brightly coloured vase-like sacks, each with two necks. Sea squirts are supplied with both oxygen and food by a steady flow of water, which is drawn in through the uppermost neck, the mouth, and passes down into a large bag, the pharynx. The wall of the pharynx is perforated by many modified 'gill slits', and the filaments that separate them are hollow and are filled with blood. The water passes through the gill slits into an outer chamber, the atrium, and out via the lower neck. The water current is maintained by the beating of numerous cilia covering the pharyngeal basket. Food particles brought in by the ciliary currents are wound into a string of mucus that passes back into a short oesophagus; from there it goes to the stomach, with associated digestive gland, then the intestine, and finally out through the anus, which is conveniently situated near the exit of the atrium.

The blood of the sea squirts is very special and comes in a range of colours, due to the presence of compounds of the rare metal vanadium. Their nervous system is relatively simple, and they have no organs of excretion, so apart from their special 'blue' blood they have nothing to suggest that they are 'high class' creatures.

The clue as to their real identity is found early on in the life cycle, in the larva. The fertilised egg develops into a larva approximately eight millimetres long and known as an 'ascidian tadpole'. Its tail is about four times as long as its trunk, and contains a stiff rod, the notochord, and a hollow nerve cord with a band of muscle on either side. So here is the answer: the notochord, forerunner of the vertebral column, and the hollow nerve cord, a ghost of the central nervous system, or spinal cord, yet to come. The free life of the 'tadpole' is only very short, but those are the accoutrements of that 'freedom'. Soon the

mouth

atrial syphon

gill slits

anus

branchial chamber

heart

tunic

A sea squirt, an animal that provides an important link in the evolutionary tree.

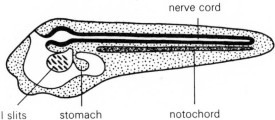

nerve cord

gill slits   stomach   notochord

Ascidian tadpole, the larva of the sea squirt.

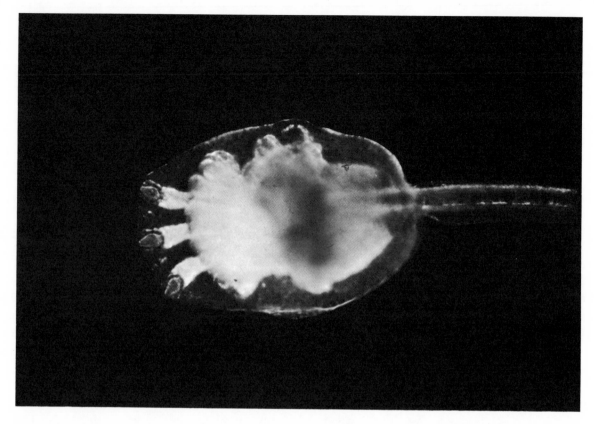

Get ahead, get a notochord. Close-up of the head of an ascidian tadpole, the larva of the sea squirt.

tadpole, unable to feed itself, settles down and loses all vestiges of its advancement to become just another sea squirt.

The *Tunicata*, the sub-phylum to which sea squirts belong, really do come in all shapes and sizes, from the solitary to the ultra complex colonies that are able to regenerate and that may reach the immense size of ten metres. The majority of them are sedentary but some, with the wonderful name of salps, are pelagic and consist of transparent tubes or rings with small coloured nuclei that represent the positions of the individuals. In the salps the egg is retained inside the parent where it grows into a tadpole nurtured through a placenta.

The tunicates are just one of a number of animal groups possessing the rudiments of the vertebrate plan. Lacking a true backbone they cannot in the strict sense be called vertebrates. They are therefore given a place of their own and are called the protochordates, but in acknowledgement to their link with 'higher' things they are grouped with the vertebrates into the phylum *Chordata*—for they have struck their own cord of success.

As well as the tunicates, the protochordates include two other groups of animals which, although they can be present in the sea in abundance, are often overlooked because of their size and habitat. One of them, the acorn worms, either burrows in the mud of shallow water or is found living in a tubular house at considerable depths in the sea. Each burrowing worm consists of a proboscis that sits in a cup-like collar (hence the name of acorn worms), with the mouth at the base of the proboscis. This is no ordinary proboscis: it contains the simple heart and excretory organ supported by the short hollow notochord. The collar likewise contains

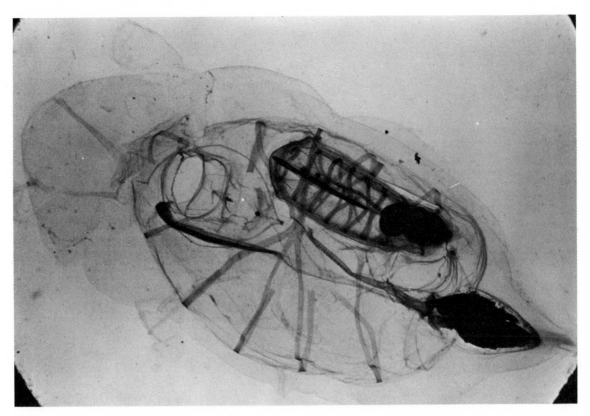

Thalia democratica, a colonial salp together with an embryo.

the hollow dorsal nerve cord, and both proboscis and collar are used for burrowing. Behind the collar is an elongate trunk perforated by gill slits and tapering to the terminal anus. The tube-living forms are greatly modified, the short body being bent round so that the anus also protrudes from the mouth of the tube.

Apart from the peculiar formation of their mesoderm, the other feature linking the protochordates and the echinoderms is their larva, which is so like that of the sea cucumbers that for a long time even the best taxonomists of the day were fooled.

The third member of this exciting group is amphioxus (a name simply meaning tapering at both ends), which looks like a small fish. Although it can swim actively it spends most of its time partly buried in the sand, and probably

due to this sedentary existence it lacks eyes. Throughout its life, amphioxus has three important attributes: a long strong notochord; a tubular nerve cord running the length of the body just underneath its fin; and many pairs of gill slits through which water, taken in through the mouth, is expelled from the body. These are the very three attributes that the German biologist Ernst Haekel pronounced in 1874 to be the main characteristics linking all the higher animals into the phylum *Chordata.* Yes! all the vertebrates, even man, at some stage in their development possess all three. However in most cases they are but transient phases, appearing, only to be lost, almost as if the developing animal were re-enacting its early evolution. So in the same way that study of the larvae gives us clues to relationships between the animals without backbones, so does study of the embryology of the vertebrates help unravel the story of evolution.

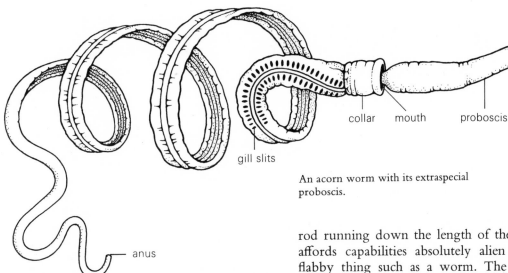

An acorn worm with its extraspecial proboscis.

The 'fishiness' of amphioxus is heightened not only by its flavour (it is akin to whitebait) but also by its long dorsal fin which is supported by gelatinous fin rays. Although its feeding mechanism is fairly basic, its gut is very like the vertebrates'. It has a true liver formed from and opening into the digestive tract, and the blood, which picks up digested food from the capillaries surrounding the intestine, passes through the liver before returning to the general circulation. This condition of a hepatic loop is met only within the vertebrates. Thus the only other features lacking from the structure of amphioxus are a vertebral column and a brain box.

Watching one of these animals swimming it is possible to see the advantages of a semi-stiff rod running down the length of the body—it affords capabilities absolutely alien to a soft flabby thing such as a worm. The blocks of muscle working against the axis allow efficient and prolonged swimming, including those rapid darting movements so characteristic of fish. Its least advanced feature is its feeding, which is basically ciliary and not unlike that of a sea squirt. The cilia are located on the bars separating the numerous gill slits, and their food is mainly protists. In contrast, in most true fish the gill number is very much smaller and the first gill arch is modified to form the jaws. Also, very few fish are suspension feeders.

In the June 1972 issue of the journal of the Linnean Society of London there appeared the first description of a new organism which was given the name of *Reticulocarpos hunusi gen et sp. nov*, the latter meaning that it was both a new species and a new genus. The Linnean Society was founded in 1788 and is named after the

opposite
An acorn worm *Glossobalanus sarniensis* burrowing into shell gravel.

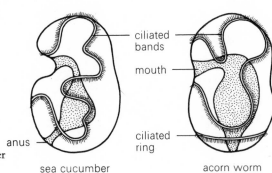

The similarities between the larva of a sea cucumber and that of an acorn worm.

sea cucumber   acorn worm

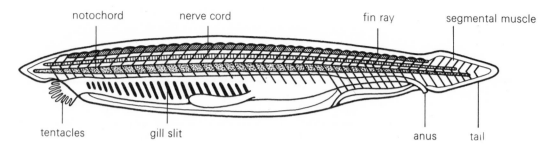

notochord   nerve cord   fin ray   segmental muscle

tentacles   gill slit   anus   tail

Amphioxus, the shape of things to come.

Swedish naturalist Carolus Linnaeus, creator of the binomial system of nomenclature for plants and animals, so its journal is the ideal place in which to publish the description of a new organism.

Lancelets or amphioxus, showing fin with rays and muscle blocks, and segmentally arranged gonads on lower side.

The fossils on which the description is based were from siliceous nodules collected from the fields of Sarka, which now lie beneath the bricks and mortar of the suburbs of Prague. The ten fossils had lain for many years in the Prague museum, awaiting detailed study. The soils of the fields of Sarka are, or rather were, made from

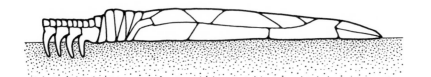

Reticulocarpos hanusi (side view)
about to take off.

black shales that had originally been laid down in a large marine embayment around 455 million years ago.

Reticulocarpos is a member of a group of extinct animals that appear to form a link between the spiny-skinned animals, the echinoderms, and the chordates, the animals with 'backbones'. All the species so far described in this group consist of two distinct parts, a tail and a body, and all appear to have lived a semi-sedentary life, lying flat on the sea floor. Their link with the sea urchins revolves around the fact that each separate plate of their external armour is made of a single, large crystal of calcite, a characteristic feature of the echinoderms. Features linking them with the chordates include the possession of gill slits, post-anal tail, notochord, dorsal nerve cord, and regular blocks of muscle.

Most exciting of all, from the fossil evidence it is possible not only to construct a detailed picture of the animal itself but of its habitat. *Reticulocarpos* lived on soft mud and its whole structure shows adaptations to overcome the problem of sinking below the mud surface. Its calcite plating was especially light to cut down the total weight of the animal, and its body was very much flattened to provide maximum surface area. There is however no doubt that it lived a pretty precarious existence poised on the mud surface, but there is also little doubt that these mud-living adaptations were equally valid for a more active swimming mode of life. With a flattened body, light skeleton, long tail, and a rod-like notochord against which the strong muscles could act, anything was possible.

Thus the first chordates were poised, ready to take off in a big way as the tadpole of the sea squirts ably demonstrates, and almost as if to round off the story, there are other members of this weird group of inbetween animals that look exactly like an ascidian tadpole with armour plating. There is still much debate concerning the exact taxonomic position of *Reticulocarpos* and its kin. Some say they are echinoderms, others that they are chordates; but all agree that they are an evolutionary step in the right direction.

The birds of the sea
fish

Just as amphioxus may be placed at the top of the protochordates, the lampreys and hagfish must be placed at the bottom of the fish world. They have no jaws, the mouth being formed into a sucker that allows them to attach to stones or to other fish. The notochord is retained throughout life and is strengthened by plates of cartilage which also include the hollow nerve cord. These represent the first vestiges of a true backbone. Feeding in the adult lamprey is by means of a toothed rasping tongue, and although in adult life they are either scavengers or parasites, in the larval stage their diet and feeding mechanism is like amphioxus, except that the water is pumped in and out by means of muscular contractions of the pharynx.

So the lampreys introduce the great vertebrates, those animals that, due to their wonderful

skeleton, well-developed brain, and efficient system of organs, muscles, glands, and kidneys, have obtained a marked though spurious independence of their environment. One of the vertebrate groups is the fish. They are the birds of the sea, reaping the potential of our inner three-dimensional space, able to move of their own free will through the restless waters.

Once under the surface, both of the enormous literature and of the sea, the fish are among the most fascinating products of evolution. The more one learns, the more amazing become the similarities between the fish of the sea and the birds of the air. Both enjoy the freedom of real three-dimensional movement, and both include among their members groups that migrate, build nests, lay eggs, defend territories, perform ritual mating dances, feed their young, and

Head end of a hagfish *Myxine glutinosa* together with its slime secretions.

opposite
No jaws but what a sucker!
Business end of a sea lamprey
Petromyzon marinus.

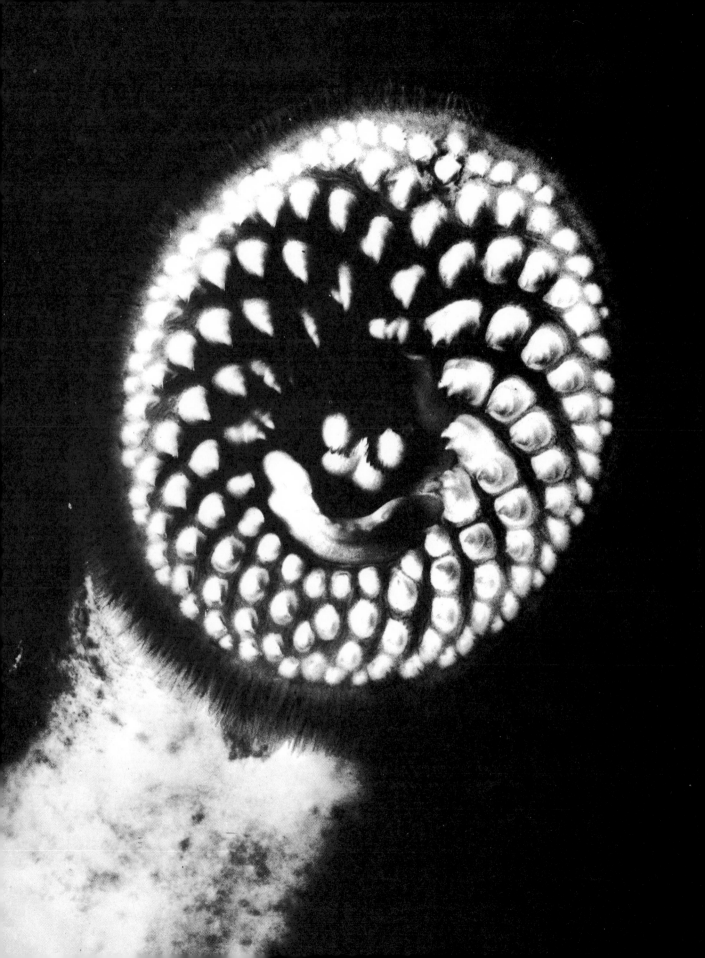

Trisopterus luscus, the bib, in water showing the dorsal, pectoral, ventral, pelvic and tail fins. The ordered rows of scales and the lateral line running from the front to the back of the body are conspicuous. The lateral line marks the position of organs that sense the fish's environment.

congregate in great numbers. In addition, some birds 'fly' under water, and some fish make short 'flights' into the air. Thus looking at one type is almost like looking at a mirror image of the other, so perhaps it would be useful to compare the problems of the birds with those of their fishy counterparts.

Archimedes found his answer in the bathtub by displacing his own volume of water. If that volume of water had weighed exactly the same as Archimedes himself, he would have floated in the bath, rather than sinking (as we must presume he did). Buoyancy is a function of weight and volume. Isambard Kingdom Brunel knew this and astounded everyone who did not by starting to build an iron ship. The same is true of air but as air is eight hundred times less dense than water, you would have to be both very light and very large to be buoyant in air. Count Zeppelin saw that when he designed his airship.

Birds must therefore expend energy in lifting themselves and holding themselves up against the pull of gravity. However, once at altitude it takes little effort to fly through the thin air. In contrast, fish expend little or no energy in overcoming the force of gravity; most can float suspended in the buoyant liquid. They must, however, expend considerable energy to force their way through it. It is for this reason that fish move so slowly. A darting trout, fast as he may seem, is moving at only about 8 kph and he is unable to keep up that speed for a long time. On the other hand, a migrating bird can keep up a flapping 60 kph for many hours on end. However, as soon as the bird stops flapping its wings it will begin to fall to earth, whereas a fish can lie at rest in mid-water.

This is not true of all fish. It is in fact only true of those more advanced species that possess control reservoirs of air that help to keep their weight and volume in the right proportion giving them the advantage of neutral buoyancy or weightlessness under water. All fish not possess-

ing such reservoirs, aptly called swim bladders, must make a physical effort to keep up at the top, even in the world of fish.

The one way in which fish do differ markedly from birds is in their diversity. The sum total of the world's bird species is 8500, compared with the 20,000 fish species that swim the oceans. But what exactly makes a fish a fish? By definition, they are aquatic, cold blooded, gill-breathing vertebrates that have fins supported by an internal skeleton of rods.

If you really want to understand the problems of a fish elbowing its way through water, then the best thing to do would be to take a powerboat out for a spin—preferably one fitted with an outboard motor. The noise of the engine, the great flurry of foam, and the rate at which the fuel gauge falls are proof enough of the power that is required. Now shut off that power and note how quickly the boat is dragged to a halt.

As the boat (or fish) moves forward, it displaces its own weight (volume) of water and this requires energy. It also leaves a space in the water

A conger eel, eeling along.

behind the moving object that is filled in by water flowing from the sides, and this creates drag. With the throttle full open the bow of the boat rides high, the stern sinking into the cavity created behind the boat. As soon as the throttle is closed the depression rapidly fills, dragging on the boat and causing it to come to a rapid halt.

Evolution has overcome a few of these problems by producing fish perfectly shaped for the job. However, a glance at the slab in even a modern fishmonger's shop will show that the 'answers' come in very varying forms so there must be many different requirements in the sea, each of which has its particular specification. Two such extremes in form and function are the long thin eel and the not so long and not so thin blue-fin tunny—shaped like . . . well, a blue-fin tunny!

The eel's body is certainly perfectly adapted for navigation in tight corners and through small holes. Also, considering that its migration takes it 6500 kilometres from its European fresh waters to its breeding ground near the Sargasso Sea, it must be ideal for sustained travel. Eel propulsion is by way of waves of curvature that pass down the length of the body starting at the head end. Thus, as the eel moves forward its head and tail undulate from side to side, the whole thing snaking, or rather eeling, its way along. At the other extreme, the tunny (and the

The tunny, streamlined for speed.

blue-fin tunny holds the world speed record of more than 72 kph) cannot flex its short fat body. The enormous drive power comes from a similar side to side movement of the strong slim tail which is worked by the powerful body muscles.

Between these two extremes come the whole range of fish best lumped together under the term spindle-like, their shape and movement being intermediate between those of the eel and the tunny. All of them are active movers and exploit the great three-dimensional water space. Their swimming power is based on the series of rhythmic body waves brought about by an intricate musculature.

A flounder *Platichthys flesus* succumbing to the force of gravity and flattening to lie on the bottom.

When you next eat a fish, remember that the bulk of what you are eating is those great power-house propulsion units. In fact three-quarters of a tunny is muscle, and they can run to almost one tonne in weight—what a catch and what a swimmer! If the skin is removed carefully it is usually easy to see the blocks of muscles arranged in strict V or VV formation across the width of the body. Before the whole thing breaks up into those characteristic flakes, count the number of muscle segments, and do not discard the backbone until you have counted the vertebrae. The two totals should be the same because the vertebrae, together with their long extensions, are the rigid framework onto which the muscles are anchored.

A fish flattened for entering narrow openings.

Carving a juicy fish steak allows a view of the musculature in section. Here the relationship between the muscles and vertebrae is much clearer: a regular pattern of sets of muscle blocks radiating out from each vertebra. These are the muscles that produce the rhythmic flexing of the body that is the basis of the fish's power.

This is all very well for the great free swimmers, the pelagic fish that ride the open waters where speed and stamina are the necessary attributes for the 'jet set' life. In the back waters of a coral reef, or in the confines of the inshore forests of seaweeds, more delicate methods of going about the daily business are called for. A blue-fin tunny in the confines of a coral reef

would be rather like the proverbial bull in a china shop (except in this case it would be the 'bull' that would suffer the damage, not the rock-hard reef fitments).

As we have seen, fish by definition have fins. Anyone who has watched a fish swimming in an aquarium will know that the fins are not only used for steering and braking but in many cases they are active agents of thrust. In the same way, anyone who has watched spellbound at the delicate movements of ailerons and flaps as an aeroplane comes in to land, will appreciate the delicate coordinated control necessary to en-

opposite
The lion or dragon fish which lives among the coral reefs. The long dorsal fin spines are poisonous.

Easy to tread on. The scorpion fish has poisonous spines
on its back and an incredible ability for camouflage.

sure a safe landing. Imagine the complexities of
adding the function of propulsion to these
'organs of control', yet in many of the fish this
is exactly what evolution has achieved.

The best examples are fish like the manta ray,
or devil fish which, thanks to Hans Haas and
Jacques Cousteau, have been introduced many
times into our television lounges. In the manta,
the tail has become useless as an organ of pro-

pulsion and the whole task has been taken on by
the delta-shaped pectoral fins. Careful observation, however, reveals that when swimming
fast, the manta flexes its useless tail from side to
side, thus proving its modification of the basic
spindle shape.

Rays may be conveniently divided into two
groups: the active swimmers and the less active
bottom feeders. The skates and sting rays, falling
into the latter category, spend much of their
time grubbing along the sea bed for molluscs and

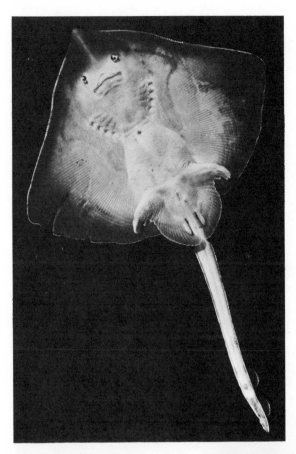

crustaceans. They can, however, rise to the occasion and lash their short, stumpy tails from side to side when in full flight, although the bulk of the work is done by the fins. These bottom dwellers are perfectly adapted for lying on the sea bed. They are negatively buoyant, which means that they do not have to expend energy in keeping down. Their eyes are set on the top of their head so that they can see anything coming at them from above, and they are camouflaged to blend in with their surroundings.

The skates and the rays are cartilaginous fish that have been flattened from top to bottom, that is, the pectoral fins form the wings. With such an arrangement, the natural place for the eyes to be is on top where they are positioned ready for action. Some of the bony fish, like the plaice and sole, have also gone in for the lying down game, but instead of being flattened from top to bottom, they are flattened from side to side. They therefore lie on one side or the other, a fact that would appear to create insurmountable problems for the eye on the underside—nothing of it! During the last stages of development, just before the young fish sinks to its flat life on the sea bed, the eye destined to

Underside of a thorn back ray showing mouth, gill slits, and the rays in the broad pectoral fins (above), and topside of a thorn back ray showing the long tail with poisonous spines (right).

Plaice in place on the bottom of the sea, with its two eyes visible on the upper side of the head (left), and the wry smile of a plaice head-on (below).

be on the underside moves round onto what will be the top side in the adult state.

These flatfish are capable of even more rapid changes than their marvellous migrating eyes: they can alter their colour to blend in with the background by rapidly changing the pigmentation of their skin. Add to this their ability of sinking in and partly burying themselves in the sand, and there you have it, a package-deal evolution kit for bottom dwellers.

In contrast, the devil rays and eagle rays live a more active life herding the plankton into their large mouths. Their eyes are mounted on the sides of their head, thus allowing for forward, sideways, and upward vision. Looking down into the deep ocean it is dark blue, looking up from deep water the light is white and blinding, therefore these great flatfish are coloured dark above and white below, thus obtaining the protection of both light worlds.

The great manta is one of the most graceful of all creatures on earth and it is impossible not to marvel at its skill in swimming, making one's own efforts seem very inadequate. They are enormous fun to be with in the water and will often let you swim alongside like some diminutive partner in a subaquatic ballet. One of the most amazing experiences I have ever had was to watch two three-metre mantas perform their ritual mating dance, turning unhurried back-ward somersaults, thus presenting their white bellies to each other just below the surface. Unfortunately the water was murky so it was impossible to view the dance from down below. So poor in fact was the visibility that the partners of the *pas de deux* lost each other and the larger one came and continued the performance to the black underside of the diving inflatable. However, evoking no response, the manta swam off, driven through the water in effortless motion by the measured, heron-like movements of its great pectorals.

The prize for what looks like effortless swimming and, come to that, navigation in tight corners, must go to the pipe fish and their close relatives the sea horses. Both have very delicate dorsal and pectoral fins that can flicker with incredible speed—each fin ray can vibrate from side to side as quickly as seventy times per second. It is this rapid but delicate movement that both drives and steers these strange fish. The pipe fish and sea horses are often found amid the fronds of seaweeds and turtle grasses, where they flicker their way through the herbage. In such a maze of vegetation, navigational precision is of prime importance. In addition, they both have delicate, elongate snouts, which makes the essential process of feeding one of accurate aim, especially in the ebbs and flows of shallow water.

A manta ray in flight.

A snake pipe fish, the ideal shape for working a way through the weed.

opposite
Sea horse among coral. The small dorsal fin provides propulsion for this stiff upright fish.

Extremely weird in appearance, although it is hard to say exactly why this is so, are the sun fish. Sun fish look like rays swimming on their sides, as they are driven along by the reciprocal motion of their dorsal and anal fins which beat out of phase. They also have a very reduced tail, and look real clowns of fish, as they scull themselves with great gusto around the coral reefs.

Whatever the method of forward propulsion, there is one very peculiar fact about fish: the bigger they are, the faster they can swim. The reason is that the increased body weight is packed into all those powerful muscles that work the larger fin system. Furthermore, unlike the birds and the mammals, fish keep on growing long after sexual maturity has been reached. So in the many fish that are not averse to cannibalism, the youngsters are always finning at a disadvantage when it comes to a race for a meal.

The most unusual use of fins is undoubtedly found in the fish that, although from a number of unrelated groups, have taken to the air. The flying fish all sport pectoral fins that have been enlarged to produce a flexible wing. The pelvic fins have also been modified in the same way but to a lesser extent, and the typical sickle-shaped tail has an extended lower lobe.

Lift off is obtained by very rapid swimming just below the surface and a final flick of the tail to launch the fish into the air. The sail fins are immediately extended and the fish glides into the wind. Once speed begins to slacken, it falls back to the surface and may either re-submerge or, if conditions are right, another quick flick will send it back into the air. A series of 'hops' may carry the fish over several hundred metres in less than twenty seconds.

Watching great shoals (or is it flocks?) of flying fish is one of the joys of a tropical cruise, and a search of the decks in the early morning will often find fish whose evening flight paths took them on a collision course. It would be nice to

think that the truly acrobatic fish fly for fun, but all the evidence points to the fact that they do it out of necessity to get away from their enemies.

Even more amazing to watch are the skipjacks which are regular performers in the shallow waters of the coral reefs. The noise they make is a staccato 'skip-flap skip-flap skip-flap', and when several thousand are skipping their way along at the same time, the noise can be heard for over two kilometres.

Flocking in birds and schooling in fish are in some ways similar phenomena, except that fish schools rarely if ever have a recognisable leader. Nevertheless, a large ordered group must afford the individuals some increased amount of protection. Although it is very easy to say that, it is more difficult to prove. Members of a school, as the name suggests, learn the tricks of the living trade much quicker than single fish that go it alone. Similarly, if all the fish were dispersed, they would occupy a much larger volume of the sea, and hence would have a far greater chance of coming into contact with predators. On the other hand, it has to be admitted that a large school of fish is much more likely to attract the attention of potential enemies. The argument usually boils down to the fact that because many highly successful types of fish, from a whole range of taxonomic and ecological groups, live in schools (one out of four of all types of fish spend at least part of their lives like this), it obviously has some very real advantage.

Schools of fish are thus very common; but all aggregations of fish are not necessarily schools. The diagnostic feature of a true school is that all its members are about the same size, and when they move they all move together. Simple aggregations rapidly disperse when disturbed; in contrast, the school comes together to present an ordered wall of fish to the predator—not that that is necessarily very effective, for large predators will still cut swaths through a clumped school.

The fact that the fish in a school are of similar size gives a clue as to how the schooling habit develops. Depending mainly on the species, a female fish can lay anything from ten to a few million eggs which will all hatch out at the same time. Thus a school may be born from one parent, the young fish keeping together. Some, like the herring, stay in school throughout their lives; others, like the rockling, join the 'educational' system only until maturity brings their brief schooling days to an end, when each goes its own, possibly more hazardous, way.

Interesting examples of protection and of schooling are to be found among the 'mouth breeders'. They get their name from the fact that the fertilised eggs are transferred to the mouth of the adult, often the male, where they are safeguarded in the well aerated waters near the gills. The male of one Atlantic 'cat' fish carries the large eggs (large that is as fish eggs go, being about 20 millimetres in diameter) secure in his mouth for about a month. Once hatched, the baby fish live in a school that still seeks the shelter of the 'good life' inside dad's mouth. It is a weird thing to watch a large fish preceded by a cloud of tiny ones that, at the approach of danger, flee in what appears to be exactly the wrong direction.

The protection game is also very much in evidence among some of the trigger fish, which lay their eggs in shallow nests scraped out of coral sand. Each pair jealously guards not only the nest but also the surrounding territory from any intruders. Some of my most painful moments underwater were experienced when entering the territory of a tiny trigger fish. He must have learned that attack is the best means of defence, for he was certainly very quick on the draw!

Each fish is thus adapted through its evolution, and some through their social finishing schools, to a particular function in the community of the sea. Apart from its skeleton, be it bony or cartilaginous, and the strong musculature, the form of a fish is to some extent dependent on its outer covering. If anything goes wrong with a fish's skin then it is going to have problems, for a

opposite
A school of spindle-shaped fish *Myripristis murdjan*.

overleaf
The fish with a long snout, *Forcipiger longirostris*.

slight change in form brought about by say a skin lesion or disease can have a marked effect on the way it moves through the water. It will thus be easily picked out from the crowd and become socially unacceptable and fall easy prey to all the hungry and waiting carnivores.

The skin is also of great importance in protecting the fish. Anyone who has handled a freshly caught one will know how difficult it is to hold—hence the saying 'as slippery as a fish'. For a long time it was thought that the mucus was produced to act as a lubricant, aiding the fish to slide through the water. This is of course untrue: those who have slipped on a wet floor will realise that water itself is a good lubricant. Although the slime probably guards the fish from the odd scrape against rocks, one of its main functions is against attack by fungi, bacteria, and other parasites. However the fish's main protection is its scales.

Watching a shark move in a business-like way under water (that is if you can overcome the apprehensive feeling as to its actual reason for moving), it is very apparent that its body movements are extremely stiff. If you are unlucky enough to brush against one or, worse still, get bitten by one, you will discover the reason for the stiffness. The shark is covered with very sharp 'scales' that are in fact not scales at all but modified teeth, and are therefore more properly called denticles. So the sharks are tough customers and due to their semi-rigid latticework of teeth, are stiff customers as well.

Such a rigid outer covering could not possibly allow the flexing of the body of the ray for example. Indeed, they only have an armour of denticles in the head and tail regions; and the most highly developed of the flappers such as the manta, where mobility of form is of extreme importance, have none at all, their skin being

overleaf
A school of big eyes.

opposite
The aggressively territorial clown trigger fish.

Tipping the scales, the orange-striped trigger fish.

completely naked. However, there are a few rays that do have a stiff outer covering of denticles, but because of this they have to swim rather like sharks do, by means of the undulations or, more accurately, the thrashing of their tails. The guitar fish and saw fish fall into this class.

It is among the less heavily armoured range of fish, especially those bottom dwellers incapable of rapid 'flight', that some really extraordinary forms of defence are found. These fish are called thorn backs and sting rays, and their defensive sharp thorns are placed in strategic array along their backs and tails. They are in fact modified, greatly enlarged denticles, and some are grooved in such a way as to collect venom, which is secreted by poison glands embedded in the skin near their base.

The electric rays need no such protection and their skin is naked. The great muscles that should power the pectoral fins have been modified to generate another sort of power, electricity, and these fish pack a punch that can only be described as shocking! In the famous Aquario at the Naples marine station, there is usually a small tank with an electric ray on display. The notice says that visitors may handle it—at their own risk.

Head of a little white shark.

Big head (not that I'd tell him) of a great white shark.

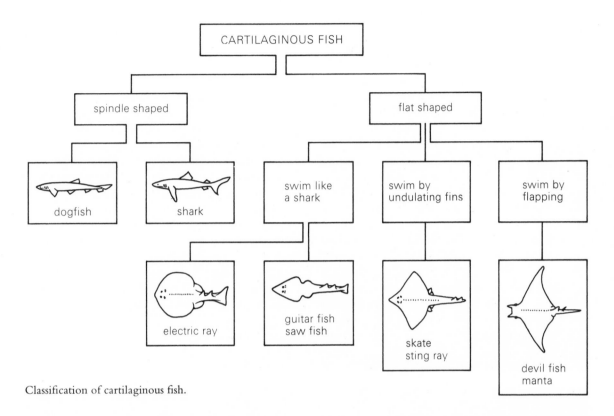

Classification of cartilaginous fish.

In contrast to the cartilaginous variety, fish that possess a strong bony skeleton have an outer covering of true scales. These are thin, bony plates that are much more flexible than the stiff denticles of the sharks. Nevertheless their presence has a profound effect on the form and function of the fish, and so again there is a whole range of scalyness. Many eels have no scales at all, which must be perfect for their snake-like swimming movements but very painful in among sharp rocks. At the other extreme are the box and puffer fish, with their veritable suits of armour. In these inhabitants of the coral reef, the skin has to be extra tough and it is well supplied with bony plates and spines or, as in the trunk fish, a bony box that encloses all but the tail, thus affording complete protection.

The best way to understand at least some of the diversity of scaling is to look at quite an ordinary fish, say a herring. The head and trunk, the least supple parts of the fish, are covered with large scales that overlap each other, giving a rigid structure. Towards the flexible tail there is a gradual 'scaling down' in size, with less and less overlapping. By bending the tail from side to side it is easy to see that they control the rigidity both by their size and the degree of overlap. The scales thus provide the required protection and yet help to maintain the form needed for a life under water, however active or passive.

It would of course be impossible to leave the subject of form and function in the fish without mentioning the most famous of all, the coelacanth. From the fossil record, it is evident that in Devonian times (about 350 million years ago) there abounded a group of fish that now go under the common name of tassel fins. There is evidence to support the idea that it was the tassel fins that first moved the evolution of animals with backbones out from the water onto dry land. Their special feature was that the anal paired fins consisted of a lobed structure supported by an internal skeleton, on the end of which the fin rays were set. If the fins were not true legs, they were certainly a step in the right direction—exactly what was needed for the evolution of a successful land animal.

This old and diverse group of fish appeared to pay the price for such evolutionary innovation and were thought to have died out about 70 million years ago. Therefore imagine the excitement of the scientific world when one was caught in a trawl off the mouth of the Chalunna River in South East Africa in 1938. Since this magical day, others have been caught on long lines, mainly around the Comoro Archipelago in the Indian Ocean. So now we know that the tassel fins are highly mobile and could indeed act like stubby legs, scrabbling over the bottom. Here was a real, live fossil that had survived for at least 350 million years, filed for life in an environment that has remained virtually unaltered throughout evolutionary time, the mid-waters of the tropical seas.

There is of course one environment, and it is by far the largest on earth, that has altered even less, and that is the great abyssal depths of the oceans. Here, in the eternal darkness of deep water, the physical environment rarely changes, except for the odd volcanic eruption. Every season is the same; the only factor that varies is the abundance of food that filters down from the plankton soup above. Perhaps it is because of this lack of change that here the world's fish are at their most diverse.

Just as the tropical lands, which have enjoyed the same climate throughout recent evolution, have the richest and most varied flora and fauna, so it is that in the tropical deep waters (below 250 metres), ten times as many sorts of fish are found as in the lighted waters of the surface. Among them are surely some of the weirdest creatures on earth: gulper eels, whose bodies consist of little more than a gigantic mouth and a

Old fourlegs, the coelacanth.

Monsters of the deep: 'have form will function', even in the total darkness of inner space.

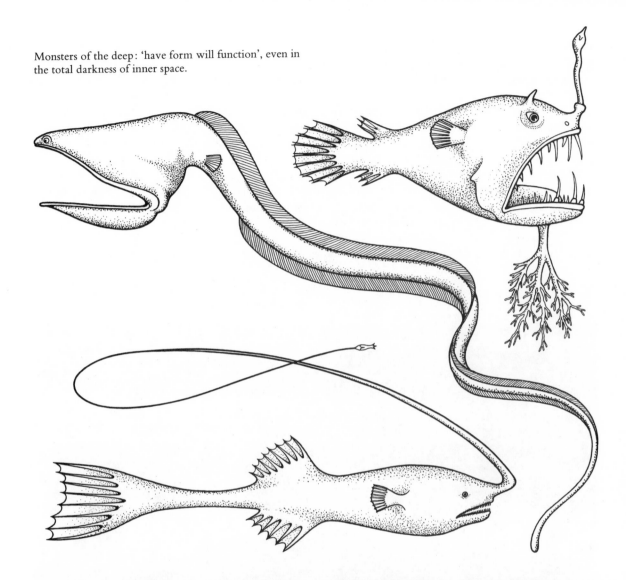

long tail, ideal for eating the occasional meal that may come their way in the darkness; lady angler fish, which have their diminutive mate in constant tow, living parasitically attached to their heads; and a host of fish that bear lights and lanterns by which they recognise their prey, predators, and kin. Most exotic of these is the deep sea angler which fishes with a luminous lure that dangles temptingly just in front of an accommodating, albeit ugly, grin. One rather fascinating fact is that the bulk of the fish that have evolved light organs live not at the greatest depths, but in mid-water, well below the zone of photosynthesis yet within the reach of the wan rays of the sun. Some of them also make a nightly migration up into the shallower water where, although less diverse, the permanent residents are more plentiful.

Strangely enough, many of the fish that have taken to living in caves are blind, their eyes being much reduced and functionless. These cave dwellers must therefore rely on touch and sound when getting around. In contrast, the eyes of most deep water fish are not fully functional but are much enlarged, as if to be able to make use of every single ray of light that comes their way.

Light from the outside world enters the fish's eye through an almost spherical lens, hence their characteristic bulging look. The lens has a high refractive index of 1.65. This means that it can 'bend' the light entering it, thus widening its field of vision. Water also refracts light, but to a lesser extent (its refractive index is 1.33) and that is why a stick appears bent when part of it is held below the water surface. The image is focused by the fish's lens onto the retina at the back of the eye. The retina consists of packets of elongate cells, called rods and cones, containing a special photosensitive pigment, that is, one whose chemical nature is altered when light is shone on it. This chemical change is instantly turned into a nervous impulse that the brain translates as sight.

The rods of the fish's retina contain a rose coloured pigment called visual purple, which responds to quite low intensities of illumination, while the cones only respond to bright light and are responsible for colour vision. Sharpness of vision is dependent on the grain of the retina, that is, the number of rods and cones per unit area. For example, one square millimetre of an eel's retina packs about three million rods—more than enough for any sun-loving fish. Before the eel migrates on its long journey

The European eel with gold in its eyes ready for migration.

opposite
A mess of elvers: young European eels on their way up river. Gills, heart, and backbone are clearly visible within their transparent bodies.

through the deep dark water, its eyes enlarge markedly and at the same time the surfeit of rods is charged with a new pigment called visual gold. This pigment is found in the eyes of all deep water fish, and it allows life in the darkness to continue, ordered and governed by the flashing photophores.

Certain of the deep water angler fish make excellent use of this fact. They have a 'fishing rod' complete with luminous lure and, in front of the lure, each species has a special array of skin flaps. These hang down like a lampshade, partly obscuring the light, and due to their highly distinctive pattern they aid identification of both friend and foe. The lure itself is just as wonderful, consisting of a blob of tissue, well supplied with blood vessels and in whose skin are embedded many bacteria. When the blood supply is switched on, so is the lure, but it is the bacteria that produce the eerie light. What is more, it is light of exactly the right wavelength to be seen by all that visual gold waiting hidden in the dark. This must surely be one of the most extreme cases of mutual help between organisms: the one a bacterium, the other a vertebrate, each in its own way evolved to be fit in this deep, dark environment.

A few years ago, while I was sitting about 30 metres below the surface of the sea on a rock platform that forms part of the eastern end of the coral atoll of Egmont in the Central Indian Ocean, I was made aware of two phenomena that were to greatly influence my basic ways of thinking about evolution.

The first was quite simple but extremely interesting. The expedition had been in the field for about six weeks and most of us were suffering from what are best described as 'coral sores'. It is impossible to work underwater in the world of the coral reef without getting cut and scraped and, with continual immersion in water, these small wounds do not heal and often become seats of infection, producing spreading sores. The cure is either to stop diving and keep the cuts clean and dry—not an easy thing to do when you are a keen diver and live on a coral reef—or to take a large dose of penicillin.

So there I was sitting with my sores (trying to put off the evil day of injecting myself), making notes on the corals that grew in abundance around me. Suddenly I became aware of gentle movements around my legs and, glancing down, saw a number of small fish busy at my coral sores. They were each about eight centimetres long and were easily identifiable as cleaner wrasse—I had inadvertently jumped the queue at a 'cleaning station'. The larger fish, groupers and sharks, which had peacefully been waiting their turn, had been overtaken by this bubble-blowing intruder who was now being 'looked after' in their place.

Not only had I entered the fascinating society of the reef fish, I had become part of it, accepted by the little wrasse who were busy feeding themselves on the dead tissue of my legs and, as was later found out, helping to effect a painless cure. At the cleaning station each fish follows the rules, and enmity between fish is, at least in part, forgotten. In some form of social truce the predator will wait to be cleaned in the same queue as its prey. After having observed this, I began to see the maelstrom of fish in a new light: not as a group of individuals struggling towards fitness, but each as part of an integrated living system that works to a stringent set of rules and within which everything has a special role to play.

The second observation was less easily explained. Thirty metres down on the reef front it is a blue world. The water column that refracts the light down from the surface has filtered out much of the intensity of the tropical sun and with it have gone all the other wavelengths of light that make up the visible spectrum. At this depth the reef and its cascades of fish are a mosaic of blues, with variations of greys and blacks—a forever moving, monochrome-washed world. Yet many of the fish and the corals themselves are brightly coloured: reds, oranges, yellows, and greens are all there but never seen.

A black grouper waiting at a cleaning station.

Catch a fish and take it to the surface and, as you ascend, the fish gradually changes colour as if some invisible artist were adding paint from his palette. Take the fish down again and it changes back to its grey-blue monochrome. Even more exciting is to wield a powerful light at depth; it is just like waving a magic wand. Wherever the light beam falls suddenly comes alive in full technicolour. A few fish are frightened by the approach of the light, while others crowd in as if basking in their own glorious colours for the very first time.

The pigments are certainly there but without light of the correct wavelength they cannot be seen. Some of these fish do move into shallow water, so that their coloration may well have a function to play in recognition or warning. But the colours of the fish and the deep water corals that never move up through the euphotic column are never revealed. Here in their deep, blue habitat, the colours themselves are of no consequence; it is the patterns they create that have functional significance. So it was at the same time that I realised the importance not only of the fitness of the fish in relation to each other, but also of their fitness in relation to the very special environment in which they have evolved to live.

opposite
Now you see it now you don't: photographed in natural light (above), and with the addition of artificial light (below).

A fish *Chromis chromis* being cleaned by a shrimp.

A nice pair of lungs
lung fish

In recent years geologists and geographers have busied themselves with the problem of fitting the world together, which is in fact their way of deciding exactly how the world took itself apart. Some of the original observations that sparked off the whole, then preposterous idea of continental drift relate to the distribution of certain animals on either side of the South Atlantic. The most famous were two lung fish. *Lepidosiren* lives in the rivers of eastern South America just above the huge bulge of land that sticks out into the South Atlantic; *Protopterus* lives in similar situations in West Africa, and its distribution is centred around the Gulf of Guinea, the part that looks as if something is missing. Well something is missing, and modern research has shown that it is the bulge we now call Brazil. Take a look at an atlas and it will be apparent that Africa and South America do, more or less, fit together rather like a gigantic jigsaw puzzle. The distribution of these two fish gives added proof to the continental drift theory, and it also has the bonus of giving a date before which the drift could not have taken place.

The lung fish evidently evolved in the drying rivers of the original super-continent before it started to drift apart. When that happened, the ancestors of *Lepidosiren* went one way while those of *Protopterus* went the other, their large paired fins becoming modified into long filaments. Both retained their swim bladders which had convoluted linings and opened to the exterior, and because of this both retained the ability to breathe air. At the onset of drought their modern counterparts simply dig in and encase themselves in drying mud and by breath-

ing through a specially maintained hole, they can quite happily remain that way for more than six months.

Swim bladders were one of the principal features that brought success to the bony fish. In the majority of them, however, the swim bladder is a closed sac that has lost its respiratory function but plays a key role in floatation. There is evidence that special glands under nervous control delicately adjust the pressure within, and in certain fish the ears are connected to the swim bladder so that the whole probably acts as a depth gauge.

It would be too easy to put our two modern 'transatlantic' lung fish at the top of the fish ladder en route to the amphibians. That is not their place; indeed they have many characteristics that push them down at the bottom close to the lampreys. However in spite of this, and in addition to their air breathing abilities, they do have a number of important features in common with the frogs, newts, and toads. Thus it must have been some 300 million years ago, two related stocks of fish took to breathing air: one gave rise to the lung fish, the other to the amphibians, although the latter did not become an important group until about 40 million years later.

The first main step in the organisation of life was to shut off, safe inside a membrane, the living chemicals from the environment. That membrane had to allow both ingress and egress of certain non-living chemicals; in other words it had to be semi-permeable, and that caused a big problem. According to the laws of diffusion, if two solutions of differing concentration are

Queensland lungfish *Neoceratodus forsteri* out for an afternoon swim on the lawn.

separated by a semi-permeable membrane then there is a tendency for water to pass through the membrane to equal out the difference. This movement is called osmosis.

All vertebrates have fewer salts dissolved in their blood than are present in sea water, and so all marine fish tend to lose water in an 'attempt', as it were, to dilute the sea water outside. Unhindered, this would be an impossible situation; however, marine fish continuously expend energy in overcoming the problem. This unsatisfactory state of affairs, together with the fact that there is no proof that the sea has increased in saltiness since the evolution of the vertebrates, provides a sufficient basis for arguing that the vertebrates evolved in fresh water. Radiating out to exploit the sea, the land, and the air, the reptiles faced the problem of water loss in all three media.

A dry skin

reptiles

One of the greatest disappointments that has come with jet age travel and satellite photographs is that we now know beyond a shadow of doubt that Rider Haggard was wrong in at least one of his many predictions: there is no 'lost world', no isolated corner in which the dinosaurs still reign supreme. The giant reptiles were all gone long before man made his first appearance—which was probably very fortunate for our ancestors. Yet I have more than a sneaking suspicion that most people would like to see one in the flesh, albeit behind very strong bars. This theory is backed by the fact that probably the most popular, and also the most terrifying, of all the movie monsters is the largest of the flesh eaters, the *Tyrannosaurus rex*, king of the tyrant lizards, with his very human arms and hands.

With life on the land came one important and far reaching development: a limb terminating in five digits that could be used for gripping and grasping. This pentadactyl limb was certainly pioneered by the amphibians before being carried to a permanent life on land by the reptiles. But what was it that gave the reptiles the freedom of the land? What was it they had that the amphibians did not?

First and foremost they had, and indeed still have, a dry skin that can protect the body, especially from loss of precious water, and that possesses only a few glands. In contrast, the skin of the amphibians is highly glandular and must be kept moist at all times because it is their main organ of respiration (reptiles breathe via

The tyrant lizard *Tyrannosaurus rex* basking in the sun at Scarborough, England.

Marine iguanas basking in the sun on the Galapagos.

their lungs). Second, and just as important, the reptiles do not have to return to the water for breeding purposes. Frogs, newts, and toads lay soft eggs in the form of spawn that gains its support by floating in the water and develops into free-swimming larvae—tadpoles, newtpoles, and toadpoles respectively. Reptiles either produce self-supporting, tough-shelled eggs that hatch directly into miniature adults, or they go the whole hog and produce live young. They are without doubt animals of the land with no remaining hang-ups about the aquatic medium

Although the real giants are no more, the reptiles having had their hey-day a long, long time ago, even today they are not unimportant members of the world's fauna. Of the four surviving stocks—lizards and snakes, tortoises and turtles, crocodiles, and *Sphenodon*—all except the latter (the tuatara, now restricted to a few islands off the New Zealand coast) include forms that live in the sea. The most famous, indeed the only, marine lizard is the iguana of the Galapagos Islands in the Pacific, first described by Charles Darwin while on his epic voyage round the world in HMS *Beagle*. Recent work aboard Jacques Cousteau's equally famous ship *Calypso* has helped to prove some of the earlier observations made on this curious beast.

Marine iguanas feed on seaweed and browse along underwater seeking the choicest clumps. They can stay submerged for more than one hour and as their body tissues become depleted of oxygen, their heart slows and eventually stops 'dead' . . . but not quite. Back at the surface, after a quick gulp of air, it is 'all systems go' once more.

Their swimming is even more clumsy and laborious than their movements on land and the whole body is used to gather momentum. Apart from predators the main hazard faced by an iguana at sea is long exposure to the cold water, for however warm the equatorial waters may seem to us warm-blooded swimmers, they can, surprisingly enough, be too cold for a cold-blooded lizard. Like all reptiles, marine iguanas are absolutely dependent on the environment to maintain their bodies at working temperature. It would seem that the sea around their island home is too cold for long immersion because they hurry back from their underwater grazings to bask on the hot rocks in the tropical sun—what a life!

The turtles must have overcome at least some of the problems related to the not so warm water for they live their lives out at sea, the mature females only returning to the beach to lay their eggs. This fact is sufficient in itself to indicate that the chelonians evolved on land and not in the water. It is in the egg-laying that their main dependence on relatively high temperatures becomes apparent. For example the green turtle only nests in places where the average surface water temperature during the coldest month does not fall below 20 °C.

For a long time the turtle populations have been on the decline, even very near extinction, and it is their eggs that have been their undoing. Inside its bony box the adult turtle is safe from most predators, and even when on dry land the adult female, though more vulnerable, is still a

tough customer. She excavates her nest with the utmost care, lays the eggs, and covering them with sand, she leaves them to be incubated by the sun. So careful is the burying operation that few if any predators are likely to find her hidden horde of high-grade fat and protein. At least that was until man came on the scene. He collected turtle eggs for food and for the extraction of fats on a commercial scale. Henry Walter Bates, the famous Victorian globe-trotting naturalist, estimated that 48 million eggs were taken each year from the upper Amazon. The adults are massacred for those very Victorian, and still unfortunately in vogue, tortoiseshell knick-knacks and for the ultimate delicacy of mayoral banquets, turtle soup. Of all the depredations of the human race, the demise of the world's turtle populations ranks among the most wanton.

Even without this intervention the egg phase is very critical. If they have been laid in the wrong place or buried either too deep or too shallow they may never hatch. However this is rarely the case and some sixty warm, sandy days later the young turtles emerge and rush headlong across the alien beach to the safety of the sea. But it is a journey fraught with danger because the appearance of the first few is the signal for all the carnivores of the neighbourhood to come and take their fill of these easy pickings.

The most wonderul part of the whole process is how the baby turtles know which way to turn when they emerge from the darkness of the nest into the full glare of the tropical sun. Experiments with green turtles have shown that even if the natural horizon is replaced by the edge of a tank the turtles will still crowd to the seaward side. This will happen even if they have been displaced from the Atlantic to the Pacific coast, where the sea is of course in exactly the opposite direction. It is therefore certainly not an in-built compass that guides them to the sea. The theory is that the difference between land light and sea light acts as a beacon leading them to the safety of, and probably out into, the open ocean.

At first the young turtles are carnivores and feed on small, weak invertebrates, but as they

A young turtle with many enemies.
Chelonia mydas, a green turtle.

mature they become herbivores with a staple diet of, would you believe, turtle grass. This fact has led to a truly amazing discovery. Turtle grass only grows in relatively shallow water and the turtles therefore come inshore and appear to have specific home ranges on which they feed. Study of certain populations feeding on the Brazilian coast included extensive tagging of the animals. The tagged turtles eventually turned up on Ascension Island, some 2300 kilometres away in the South Atlantic. Their feat of navigation becomes even more fantastic when it is realised that their island target is only eight kilometres long and that, unlike birds, the turtles lack the advantage of altitude to correct any faulty navigation; they have to do it all from sea level. Both the males and females make the long journey, for fertilisation takes place in the shallows just before the female comes ashore to lay the next batch of master mariners.

The largest of the five species of marine turtle is the rare leather back. Despite the absence of a thick protective shell, it can tip the scales at more than half a tonne and measures more than two metres from stem to leathery stern.

Turtles have a phenomenal ability in staying submerged; indeed, when at rest in shallow, well-oxygenated water they have no need to rise to the surface to breathe. Of all the unlikely portions of their anatomy it is the anus, or to be more exact the cloaca, that acts as an accessory organ of respiration. The cloaca consists of two main chambers whose walls are modified to reabsorb water from both the faeces and urine. This water conservation unit is put to good use under the surface: here the same chambers are used for oxygen exchange and are efficient enough to keep the tissues of the resting turtles adequately supplied.

The case of the disappearing turtles may have a happy ending for recent research and development has turned turtle farming from a pipe dream into a practical reality. On the farm the most vulnerable phase (from the egg to the juvenile turtle able to fend for itself) can be made almost one hundred per cent safe. Once set free the juvenile turtles will of course return to the farm beach when it comes round to egg-laying time. Already farms on the Cayman Islands, which were originally called the Tortugas by Columbus because of the immense numbers of turtles, are producing a surfeit of young turtles so that some can be returned to the natural populations in an attempt to restock the empty beaches.

Of all my underwater experiences the one I remember with the least pleasure was my first encounter with a sea snake. Snakes are close relatives of the lizards, lacking only the legs but unfortunately possessing a very efficient pair of hypodermic fangs that may or may not inject a dose of lethal poison. It is probably this strange lack of limbs and the promise of poison that go to make snakes most people's number one pet hate. Until I met this one face to face underwater I prided myself that snakes did not really worry me. It was brightly coloured and its body

An olive sea snake stranded at low tide.

was flattened making it ideal for swimming in a side-winding way. As it appeared full steam out of the turtle grass by my feet, I do not know who was the most surprised, I or the snake, but we both took off in opposite directions. Sea snakes are thermophilic and will often wrap themselves around a diver to keep warm, so on reflection I was obviously very lucky—or very cold.

Crocodiles are mainly fresh water inhabitants but some do venture into the brackish medium. Indeed, the largest of all living reptiles is the estuarine crocodile, which may be as long as nine metres from the tip of its powerful tail to the end of its long thin snout. It is in the crocodiles that for the first time we find a four chambered heart in which the freshly oxygenated blood coming in from the lungs is maintained separately from the deoxygenated blood coming in from the body. The importance of this cannot be overstressed. The tissues of large active animals require oxygen and require it fast. It is thus expedient to keep the two streams completely separate and in the crocodiles evolution has almost accomplished the task, although there is still a pore, the foramen of

Panizza, between the two ventricles through which a certain amount of mixing could occur.

The body cavity of the crocodile is divided into a thoracic and an abdominal cavity by means of a diaphragm which, although itself not muscular, is worked by a special muscle that aids ventilation of the efficient lungs. Furthermore, crocodiles are in part able to control their body temperature and, as it is now held by some authorities that in all probability warm bloodedness developed in the dinosaur line that gave rise to the birds, the crocodiles are very important beasts for study. Here again is a group of animals whose numbers have in the past been decimated by man but now, thanks to active farming, they may soon be on the increase.

An adult crocodile. Note the tears!

Birds of a feather
seabirds

In a glass case, standing alone in a corner of the zoology department at the university where I work, is a great auk, one of about eighty specimens preserved in the museums of the world. These, together with about the same number of eggs, are all that is left of a once thriving population of this bird. In the seventeenth century great auks, or garefowl, were plentiful on the islands of the North Atlantic; they were how-ever flightless and thus easy prey for marauding man. Slaughtered for food, feathers, oil, and later as rarities, the last one was killed in 1844. The great auk was the first bird to be given the name of penguin long before those of the southern seas were discovered.

Today there are seventeen different sorts of penguin and all, except one, live well inside the southern hemisphere. (The exception forms a

Two jackass penguins *Spheniscus demersus.*

opposite
A penguin from Snares Island (south of New Zealand).

unique member of the unique fauna of the Galapagos Islands that straddle the equator in the Pacific.) The most famous is the emperor penguin, which raises its majestic head over a metre above the ice and snow. This remarkable bird has chosen to breed on ice in the darkness of the Antarctic winter. The single egg is cradled on the feet of the male and covered by a special pouch-like fold of his skin. Incubation lasts six weeks during which time the patient father may lose more than one-third of his body weight. As the chicks hatch, the females return well fed from the sea to help tend the young.

Penguins are gregarious birds both at sea and especially when breeding, although their continual bickering makes it impossible to use the word sociable, and their rookeries must rank among the noisiest and smelliest places on earth. So great is their social instinct that in the depths of winter certain species organise themselves into large groups, taking turns to stand at the edge to shelter those within from the icy wind.

Feathers and an ample layer of fat beneath helps them to tolerate extremely low temperatures. They are however not immune to the cold and there is good evidence to show that even when in good health they suffer from cold feet. When standing about on the ice the penguins rock back onto their heels and even go right back to sit balanced by their short stubby tails, thus keeping much of their soles clear of the ice. Their feet are highly modified both for swimming and for ice life. They lack sweat pores and so are not troubled with sweat freezing to the ice. Also, their feet are maintained at a lower temperature than their bodies by means of an ingenious heat exchanger device in their legs. The veins and arteries supplying the feet run very close together, and in some penguins actually inside one another. Thus warm blood coming down from the body is cooled by cold blood coming up from the feet, and vice versa. In this way considerable conservation of body heat is attained so one presumes that the penguins do not suffer from the terrible discomforts

Adélie penguins at Hope Bay, Antarctica.

opposite
A puffin, the clown of the seas.

we experience when our extremities are exposed to low temperatures.

With such basic modifications of both their anatomy and their physiology it is rather surprising that penguins do range into the tropics where keeping cool must pose just as much of a problem. The difficulty is overcome by another heat exchanger: underneath the flippers there are very few feathers, no fat, and tens of thousands of capillaries that bring the warm blood close to the surface. Instead of panting, penguins simply stand with their flippers outstretched fanning themselves.

As the majority of the pictures we see show penguins at the rookery it would be very easy to get the idea that they are birds of the dry land. They are not; indeed they are the most aquatic of all birds. In the water they are excellent swimmers, vying with seals and porpoises both in speed and skill. They can make long excur-

sions underwater, rising to the surface only at irregular intervals to breathe. Some, like the Adélie, are able to progress in a series of spectacular leaps that take them clear of the water. When landing, and especially when being chased, they are able to jump over two metres to end up in a skidding heap on the ice-floe.

Apart from man they also have a number of natural enemies, the most vicious being the leopard seal which lies in wait in the sea near the rookeries to wreak havoc among the swimming birds. Film brought back for the natural history unit of the BBC has shown that here is a case of an animal killing if not for sport then at least far more than it can immediately eat.

Bird watching is probably the most popular of all the branches of natural history and the world over it is two types of coastline that attract armies of dedicated binoculared ornithologists. Estuaries and mudflats are meccas

Arctic tern.

232

Giant petrel and chick on King George Island, Antarctica

for large populations of wading birds that feed on the wealth of life residing in, on, and above the soft mud. The other habitats of special interest are sea cliffs and rocky islands where countless numbers of gulls, gannets, auks, cormorants, and shags, to name but a few, nest in disagreeable array.

Many seabirds, like puffins, terns, shearwaters, and fulmars, nest on the ground and it is for this reason that they have their breeding strongholds on small oceanic islands devoid of large carnivores. In some of them, and especially the burrowing forms, their legs are too weak to support the bird on the ground and, in the absence of a clear downhill run, they find it impossible to attain lift-off. It is thus not difficult to imagine how flightless birds like the extinct dodo and the now very rare Aldabran rail evolved. Upon arrival at an island home devoid of predators and with, in addition, a plentiful supply of food, the ability to fly would be of no advantage. Indeed, owing to the enormous amount of energy required by the large active flight muscles, it could become a distinct disadvantage. Add to this the fact that while a bird is off at sea another one could nip in and grab its nesting place or even snatch its eggs, the less mobile members of the population could well find themselves at a distinct advantage over the others. This is in essence the basis of natural selection—the fittest surviving and the population in this case eventually becoming flightless.

Warm blood and a constant body temperature made possible the evolution of muscles with sufficient power to raise the birds off the ground into the non-supporting air. Flight is aided by feathers that can be preened to perfection because a smooth outer contour reduces drag and produces an aerodynamic surface. The preen gland is situated at the base of the tail and is the only gland on the surface of a bird's body. The oil it secretes is of immense importance in the preening act but is especially important in marine, 'water off a duck's back' type birds, for without a sleek water-repellent surface the feathers could become waterlogged and lose their essential property of heat insulation.

It is however the skeleton that shows the greatest modifications for flying. Birds' skeletons consist of a few hollow girders attached to specially shaped plates onto which the muscles are attached. In the best fliers of all, like the albatross, even the wing and leg bones are hollow with internal struts to make them both light and strong. It was the basic similarity between the skeletons of modern birds and those of the bipedal dinosaurs that gave the initial clue to that at first sight most unlikely link in evolution.

Flight gave the birds one great advantage over all other organisms: the possibility of transworld migration, allowing them to exploit the seasonal climates of the earth and enjoy a life of continual spring—a privilege shared only by certain millionaires. There are just under 300 different sorts of birds that may be seen at sea, and their adaptations to marine life are quite fascinating. Evolution has not as yet produced a bird with a floating egg, but who knows what is to come?

The flightless rail of Aldabra.

opposite
Black-browed albatrosses displaying (Falkland Islands).

Mermaids and their kin
mammals

As man pushed northwards in search of new resources he discovered Copper Island in the Aleutian chain in 1741. On the island he found a large population of giant animals, the adults weighing up to two tonnes. They lived an unhurried life grazing on the productive kelp beds that thrived on the rocky shore. Knowing nothing of man and his ways and having few or no natural enemies, these animals were easy prey. A bang on the head was all that was needed to stock the ship's larder with two tonnes of succulent, fresh meat, a welcome change from the salted beef and hard tack that was then the essence of life at sea. So famous was this source of food that Copper Island became a mecca for shipping in the vicinity, and only twenty-seven

opposite
An albino fur seal *Artochephalus austraus* from South Georgia, Falkland Islands.

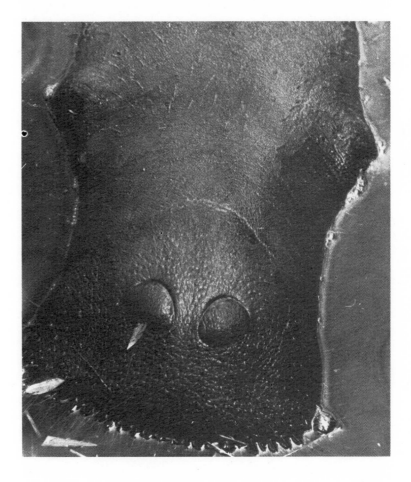

Central American manatee
Trichechus manatus showing facial hair.

years after this animal, called Steller's sea-cow, was first described, it had become extinct, wiped from the face of evolution.

Steller's sea-cow was a member of the *Sirenia*, a group of mammals that have returned from the land to graze the pastures of the oceans. They were given this name because of all the animals known to inhabit the sea they probably came closest to the mythical mermaids, the sirens that were meant to have lured many a ship onto the rocks. Although the sea-cows are mammals and suckle their young in similar fashion to a human mother, it would surely take a very drunk sailor even after many months at sea to mistake one for a comely mermaid.

Steller's sea-cow was the only representative of the sirenians that evolved to use the potential of the rich beds of seaweed that abound in the cold temperate waters. The five remaining species of sea-cow all inhabit the warmer waters of the tropics and sub-tropics. Their entries in the *Red Data Book* of endangered species published by the International Union for the Conservation of Nature read like five documents of doom. All are in danger and the reason is both the need but in most cases the greed of man. It is a disturbing fact, but of all the marine animals that have and are being hounded to extinction it is man's closest relatives, the mammals, that have suffered the most. Whether it is seals, sea lions, walruses, or whales, the story of slaughter is the same.

The largest creature that has ever lived, the blue whale or *Balaenoptera musculus,* is itself in such imminent danger of extinction that it warrants a pink page in the *Red Data Book*. Its length is 33 metres, girth 12 metres, and weight an incredible 150 tonnes. And this is the problem, for it is capable of yielding 140 barrels of oil. In blunt monetary terms, one blue whale is usually taken as equivalent to two fin, two and a half humpback, or six Sei whales, these being the main whales of commerce. It is somewhat ironical that the very feature that gave these warm-blooded creatures the freedom of the coldest seas ultimately was to lead them into such danger.

Mammals are warm blooded. They evolved on the land where their thick coat of hair and their ability to suckle, nurture, and care for their young has unquestionably given them first prize for being the fittest for life there. That such a successful group of animals should produce forms to re-invade the water is not

A white-capped hair seal proving it is a mammal.

opposite
A bull sea lion from the Galapagos.

Mother sea lion *Neophaga cineria* with pup.

Young elephant seals *Mirounga awgustirostris*.

surprising, although the modifications required for successful ocean life were very great. The most important of these was the acquisition of a layer of fat to insulate the warm body core from the cold water outside.

All living things produce heat as a by-product of their metabolism. In many it goes to waste, radiated out only to warm the environment. The mammals and birds have turned it to good use, maintaining the functional parts of their living system at their own optimum working temperature. In many land dwelling mammals it is often a case of too much heat, and it must be got rid of, for overheating is as much a problem as overcooling. The former is solved by the warm blood being brought close to the surface of the body where heat can be lost by radiation, the latter by providing a layer of insulating fat between the blood and the skin.

This is the function of the whale's blubber, and it is the oil in the blubber that has led man to

A sea otter among giant kelp. If there is reincarnation I want to be a sea otter.

slaughter these animals in such a reckless manner that the world population of certain species is, in the opinion of certain experts, below viability. In 1930 there were an estimated 40,000 blue whales; by 1962 this had been flensed down to between 1000 and 3000—a reduction that can only be regarded as pure vandalism. Their one hope is that the experts' predictions of what constitutes a viable population are wrong and that all the nations of the world will follow the directives of the International Whaling Commission so that the largest creature that has ever lived may continue to grace the oceans.

If 150 tonnes does not conjure up the right picture of this singular beast then try these statistics for size. 'The blue whale's mammoth heart pumps 9000 litres of blood around the huge body. A human child could crawl with ease through its main aorta and a full grown trout could swim comfortably through most of its major blood vessels. In the first seven days

of its life a baby blue whale, born at 2 tonnes, doubles its weight and when weaned about seven months later it weighs about 24 tonnes and has reached a length of some 14 metres.' These almost fictional facts are taken from a fascinating book published on behalf of the Whale Campaign of the New York Zoological Society whose efforts, with the right backing, may yet save the whales.

They may be conveniently divided into two groups: the baleen whales and the toothed whales. The former include the four largest, all of which feed on plankton and especially on the large shrimp-like krill. Their feeding mechanism is definitely one of filtration and the active agents are the long baleen plates that hang down from the roof of the mouth. These effectively strain the krill out of the water. The whale then simply presses its large tongue up against the palate, forcing the water out from the mouth so that the krill may be swallowed, ready salted. The toothed whales are active predators and feed on a range of flesh from fish, to squid, to (in the case of the killer variety) other whales.

Of all the research carried out on whales in recent times that of Dr Roger Payne stands out among the most interesting. Ever since whaling has been practised, whalers lying at rest in the quiet of their wooden hulled ships have reported hearing strange ethereal sounds, especially when searching for humpbacks. Humpback whales are the exhibitionists of the cetaceans. It has been known for a long time that every year a herd of humpbacks appears off the coast of Bermuda as they pause in their spring migration north. Here they have a 'whale of a time'. Their play includes spectacular leaps that take their bodies almost clear of the water to fall backwards in a deafening sheet of spray. What Dr Payne managed to prove beyond doubt was that as they move their happy way through the ocean they sing songs, well at least regular repeating patterns, some of which last as few as six, others for more than thirty minutes. He has shown that the humpbacks have several 'songs'

in their repertoire and although there is no evidence that they are communicating specific facts, there is little reason to doubt that these sounds, which under perfect conditions might travel through the water for as much as 160 kilometres, could help the members of the herd in keeping together. Will the plaintive call of the whales be used to track down the remaining members by man, or will it be allowed to rally their dwindling numbers together to restock the empty silent oceans?

The most fascinating question posed by the whales is, why are they so large? Undoubtedly it is the support provided by the aquatic medium that has made their enormous bulk possible. Their huge volume and relatively small surface area greatly reduce heat loss: the efficiency of the blubber in heat conservation is evidenced at death when the whale cools down so slowly that its insides begin to decompose with the formation of much gas. Furthermore the ratio

Crab eater seals.

of volume to surface area reduces the frictional drag on unit weight of whale moving through the water. There is just no getting away from the fact that whales are very competent swimmers, and when it comes to diving they have no equal, for they can make trips to depths probably in excess of 2000 metres.

Their modifications for this are complex. There are valves to close off the nostrils and special cartilaginous rings to keep the bronchioles of the lungs distended while under pressure. The large elastic lungs can be rapidly filled with air and when on the surface, whales normally breathe only once every few minutes, blowing out jets of water as they exhale. As well as the large lungs, their excessive blood volume acts as an oxygen store, and in addition there is a good deal of oxygen-carrying haemoglobin in the actual muscle tissue.

The modifications for swimming are no less fantastic. Large as the whale's vertebral column is, it bears no weight and is altered to form a compression strut, as in the fish, against which the muscles work the up and down movements (unlike the fish, their tails are horizontal) of the enormous fluked tail. The paddle-like flippers are modified fore limbs, yet the hind limbs have been completely lost, the tail flukes being developed from mere folds of the skin.

Here is a land animal scaled up and adapted to life underwater. Its lifeline to the air up above has to be maintained by regular trips at intervals of no more than thirty minutes in order to take great gulps of life-giving air. Apart from that vital link, the whales are masters of the aquatic medium.

In recent times studies like those of the songs of the whale have made man look closer into the brainpower of the cetaceans. Although as yet we are incommunicado with the larger of the whale clan, man has struck a relationship of mutual understanding with the dolphin, and their antics can be seen at the many dolphinaria that have sprung up along the coasts of the world. Although we may well question the use of these animals for such circus antics, especially if any hint of cruelty comes into either their training or sardine-can life style, at the same time we must remember that all our domesticated animals were once removed from the wide open range.

Much more ominous is their potential use in underwater espionage and warfare, and Hans Hass, who was among the first to bring the underwater world into our lives through the medium of television, has raised his considerable lobby to oppose such acts. It must be hoped that man will move into an era of peace where the lessons learned in the training of 007 dolphins can be turned to better use.

The southern sea otter, *Enhydra lutris nereis*, is a lovely animal that lives its peaceful life cropping the shellfish from sub-tidal waters. Of all the true marine mammals this is the only one that has learned to use a tool. In order to deal with the tough shells of the abalone, the otter selects a stone to use both as a hammer to dislodge the mollusc from the rocks, and as an anvil pillowed on its abdomen on which to smash the shell. The story of the southern sea otter rings an optimistic note of success in the whole sad story of marine mammals. Thought to be extinct in 1911, a small surviving colony was found near Monterey, California, twenty-seven years later. Since then, thanks to protection, it has been on the increase but its future is still uncertain.

The statistics books say that one man held his breath for 13 minutes thus setting not only a fantastic record but the absolute natural limit for underwater man. Despite what the adventure films would lead us to believe, it is impossible to breathe through a long tube while submerged more than a few centimetres below water. The reason is that the pressure of the water on the rib cage makes it impossible for ventilation of the lungs with air under normal pressure. Ever since the dawn of the age of invention, man has produced ingenious designs and devices for living and moving underwater. As early as 1864 a 'submarine' was successfully used in action in the American Civil War. But it was not until very much later, in 1937, that Augustus Siebe perfected a diving helmet and suit that allowed man to take his first long walk underwater, although even then his freedom was limited by the length of the hose that supplied him with air from a pump on the surface.

Many ingenious modifications followed and by 1943 Jacques Cousteau, with the help of Emile Gagnan, had perfected their aqualung. It consisted of a bottle of compressed air with a regulator valve supplying air on demand at the required pressure. All that was needed to complete the transformation was a diving mask so that Cousteau's eyes could look into the depths through a reservoir of air rather than directly into the highly refractive water. With a self-contained breathing apparatus and clear underwater vision, man took his place for the first time among the fish. Although technology is now sending the diver deeper and deeper in safety, in the shallows (down to a depth of about 60 metres) the hardware is already quite sufficient and with the right training anyone with the will to succeed can have the experience of a lifetime.

Man has neither the excess blood volume of the whale nor the low oxygen requirement and anal 'gills' of the turtle. He must therefore understand the limitations of his own physiology and psychology within the aquatic medium. He must know that while breathing at depth both oxygen and nitrogen are dissolving in his blood: the former can be used, the latter can develop into bubbles as he returns to the surface. If his return trip is too rapid then large bubbles may form. Should these be carried to a joint, it will bend causing excruciating pain, or should they reach the heart, death may result. A well-trained diver knows this and therefore carefully times his ascent in relation to the depth and length of his dive.

The bends are just one of the many problems that face man underwater and as adequate training cannot be gained only from books, practical experience in the sea is essential. Through his training the diver confers on himself the freedom of the shallow seas. He can then share the experience of the astronaut in exploring new 'landscapes' and the excitement and the responsibility of being first.

There is one marine mammal which, without the help of external navigational aids, makes its own excursions into uncharted landscapes. The animal in question is the Weddell seal, and the diving boffins would love to know a lot more about it. This hardy creature, as its name suggests, lives in the Weddell Sea which, with its average frigid temperature of $2\,^{\circ}$C, freezes over to a depth of around a metre for as much as eight months of the year. The seal lives under the ice and maintains contact with the fresh air world above through breathing holes, natural breaks in the ice. It keeps these open by means of a widely gaping mouth and strong incisor teeth, which are used to ream out any newly formed ice from underneath.

This complete dependence on a breathing hole makes the Weddell seal easy prey both for hunters and scientists alike. Gerald Kooyman is one of the latter. With the aid of a helicopter, an ice pick, and some highly sophisticated seal bugging electronics, he has learned much about the sub-ice life of the Weddell seal.

The basis of his experiments have been to create a new blow hole isolated from any others and then introduce a well bugged seal through it. It is thus faced with a new home in uncharted waters. Once released, the seal plus electronic package characteristically makes some shallow dives as if to locate its new source of air. Then, becoming more adventurous, it takes off on longer journeys lasting up to about one hour, returning to take another big gulp of air. Apart from the problem of location, the fascinating question is how does the sub-mariner know when to turn back? Does it possess some clock warning system linked up to both its metabolism and integrated with its own strength and that of the prevailing water currents? Whatever the answer is, it must be a highly sophisticated system for, experimenting with a series of strategically placed blow holes, it was found that if there was an 'unknown' hole within safe swimming distance, then the seal would carry on past the point of no return in a sort of sixth sense knowledge of the existence of this 'unknown', within-range blow hole. Sight cannot

Weddell seal all curled up.

enter in, because many of these feats of predictive navigation are carried out in the darkness of the Antarctic winter.

Furthermore, the Weddell seal has already been shown to possess an extremely advanced diving technology. During long trips beneath the ice the seal's heartbeat slows to about one-tenth of the normal rate as the blood flow is shut off to all organs except the heart and the brain, thus conserving precious oxygen. Most surprising of all, the seal can make repeated excursions to depths in excess of 600 metres, each lasting about fifteen minutes with a short rest period in between. A fascinating question is thus raised—why no problems with the bends? The answer lies in the fact that the lungs of the Weddell seal are so constructed that they collapse completely under pressure, expelling the contained air. Below 50 metres there is no air left in the seal's lungs and hence no problem of continued uptake of nitrogen, and no nitrogen means no bends. Kooyman's bugged seals prove

beyond doubt that the 'sealed' system is fully aware of the facts of life because they rarely make long excursions to depths of less than 50 metres.

As far as man is concerned there are still barriers—technological, physiological, and psychological—to be broken, but with the money and effort now being poured into inner space research, the problems should soon be solved and the third dimension of the aquatic world opened 'deep' to man. *Homo aquaticus* is the dream of Jacques Cousteau: man working alongside marine life to share the true richness of the ocean. But is it a nativity or a requiem? *Homo aquaticus* came into being in 1943 with the invention of the aqualung, and he is already proving his abilities in reaping the resources of the sea. The question is, has *Homo sapiens* evolved sufficiently to shoulder the responsibility of this new way of life?

245

potential
to man?

Some component systems

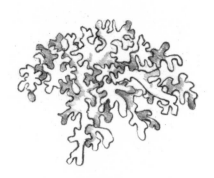

The process of evolution has slowly but surely changed the lifeless seas of 3000 million years BM (that is, before man) into those of today teeming with countless forms of life. The whole pace of life in the sea is dependent upon the amount of incident energy beaming down from the sun that is fixed by the photosynthetic plants. This is in turn controlled by two linked factors: first, the amount of light that penetrates the water, and secondly the amount of certain key plant nutrients present in the euphotic column in a form available to plant growth. The greater the available nutrients, the greater is the potential floating crop of plant plankton, but as the thickness of the plankton soup increases so the depth to which light can penetrate into the sea is reduced, and thus the plankton population is controlled.

Over much of the open ocean the actual floating crop is far below the maximum possible, being held down by the low levels of available nutrients, especially phosphorus. The reason for the low levels is that once any nutrients have been entrained into the food web they soon come under the action of the force of gravity which carries all dead planktonic organisms down through the water column. Although the nutrients may be recycled many times en route, their final destination is either resolution at depth or the detrital ooze which covers the ocean floor, softening the contours of the seascapes of the deeps.

When Jacques Piccard made his descent in the bathyscaphe *Trieste* to the record depth of 10,900 metres off Guam in the Pacific he described a waste of snuff-coloured ooze. He also saw two living organisms: a flatfish and a beautiful red shrimp. Piccard came back to tell his tale, but for most things falling into the abyss it is a one-way journey. The great thickness of the ooze carpet is sufficient to prove that very little returns to the surface. It is therefore a vast resource of nutrients locked up in the eternal darkness. Yet life does go on in the waters above the abyss and some of the component systems are very productive.

The Sargasso Sea

The very deep did rot: O Christ!
That ever this should be!
Yea, slimy things did crawl with legs
Upon the slimy sea.

Thus spake Coleridge's ancient mariner concerning what was at one time one of the most feared parts of the open oceans, the Sargasso Sea. The presence of the great masses of weed floating on the surface of the North Atlantic Ocean was first noted by no less a person than Christopher Columbus on his epic voyage of 1492. This started a whole series of stories that led the average landlubber, and come to that the average mariner, to imagine an enormous whirlpool-like area covered with a mass of rotting weed that would entangle any becalmed ship in a fetid trap.

In reality, nothing could be further from the truth. The ordered rows of seaweed laid out by the interplay of wind and current are sparse enough to allow even the smallest ship an unhindered passage. However it is easy to understand the consternation among the members of Columbus's crew. They were experienced men of the sea and knew that large brown seaweeds only grow in shallow water on rocks, so all the indications were that somewhere quite near lay an alien, uncharted reef. What they did not know was that the sea was more than five kilometres deep under the keel of the *Santa Maria*. The further they travelled across the seaweed sea the more they must have puzzled where it was all coming from—a problem that has foxed the naturalists ever since.

The seaweed *Sargassum* gets its name from the Portuguese *sargaço*, meaning grape, because the small air bladders that give this intricately branched weed its floating power look very like small yellow grapes. As to the origin of the weed itself, for a long time fingers were pointed at various parts of the world where the rocky coastline could represent its home anchorage.

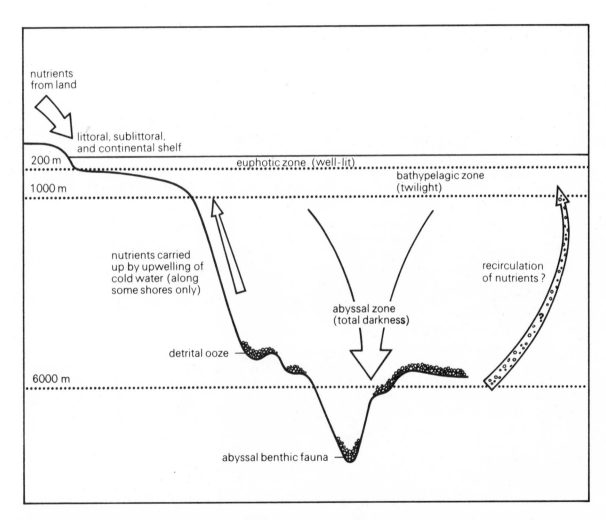

A working ocean. The littoral (including estuaries), sublittoral, and continental shelf areas can be very productive, netting more than 30 grams/m²/day (the average figure for inshore waters is 2.5 grams/m²/day), while the euphotic zone in the open ocean has a low productivity, averaging 1 gram/m²/day, held down mainly by the deficiency of nutrients. Both the bathypelagic and abyssal zone food webs are detritus based. Little or nothing is known concerning the recirculation of nutrients in open ocean areas away from upwellings. If it happens at all it is probably a very slow process.

However, once rough estimates had shown there to be at least seven million tonnes of it free on the high sea it was realised that none of the supposed breeding grounds could produce that much. Furthermore, two of the most abundant species of floating *Sargassum* have never been found attached, thus the mystery deepened. Today there is every reason to believe that the weeds of the Sargasso Sea live in the open ocean, multiplying by vegetative reproduction, that is,

splitting up into new individuals as they grow. This leaves one major question unanswered: why do they stay there?

Next time you empty your bath, watch the water as it descends the plug-hole. If you live in the northern hemisphere it will go down in a clockwise direction, if you live in the southern hemisphere the opposite will hold true, while on the equator it will go straight down. The experiment certainly does not work everytime,

A windrow of *Sargassum* weed.

opposite
A young trigger fish about to disappear into *Sargassum*.

but if the bath is perfectly smooth and the water absolutely still to begin with, then it should follow the rule.

The reason for this is that as the earth spins on its axis it sets up currents in bodies of water, however small or large they may be. Because the circumference of the earth is greater at the equator than it is say at the Arctic and Antarctic circles, a given point on the equator must travel much faster than one at either a higher or lower latitude. This differential sets the water in motion producing a basic pattern of ocean (and bathtub) currents. However the sea is neither smooth nor still because its surface is buffeted by wind, and so the actual currents are much more complex than the basic north-south, clockwise-anticlockwise ones that theory says should exist.

Sweeping north eastwards across the Atlantic are the Gulf Stream and North Atlantic Drift which form the northern boundary of the Sargasso Sea. This vast and complex stream of warm water helps to accentuate a clockwise 'whirlpool' effect that herds the floating *Sargassum* into a roughly elliptical area covering five million square kilometres, and it has been shown that the water is piled up nearly a metre higher in the middle than it is at the edge.

Seven million tonnes divided into five million square kilometres means that the dreaded *Sargassum* is pretty thin on the ground, and when it comes to biological productivity the huge Sargasso Sea is down at the bottom of the class. Final proof of this rather unexpected fact came only when Einar Steemann Nielsen, a Swedish oceanographer, measured the productivity of the world's oceanic plankton.

His method was simple: two containers, one transparent, the other opaque, were filled with water from the euphotic column. Radioactively labelled carbon dioxide was then added to each and the containers were re-suspended in the sea where the plankton got on with their task of photosynthesising in the light bottle (using up carbon dioxide), and of respiring in the dark bottle (giving out carbon dioxide). The two bottles were then retrieved and the plankton was collected from each so that the amount of radioactive carbon dioxide fixed by the plankton could be measured using a sophisticated Geiger counter. Simple addition and subtraction gave an estimate of the photosynthetic productivity of the sea at that point. Nielsen thus showed that the actual photosynthetic potential of the Sargasso Sea is only about one-third that of average oceanic water. Whatever the stories say, it is certainly not a seething mass of life abounding with sea snakes and fish; it is, as seas go, a desert and is now reckoned to be the clearest, purest, and biologically speaking the poorest stretch of all the oceans.

The anchoveta

Not all the oceans are deserts and yet for the most part they are of very low production potential, averaging less than one gram of dry matter per square metre per day. A similar average for land based vegetation is about three grams per square metre per day. The factor limiting open ocean production is without doubt the scarcity of nutrients in the euphotic zone. There are however places where currents sweep up from the depths bringing with them nutrients to fertilise the lighted water. Most famous of these is the upwelling off the coast of Peru, where prevailing winds drag the surface waters offshore. This is replaced by water flowing up the face of the continental shelf, which thus enriches the coastal waters. Here the anchoveta, the most productive fisheries in the world, are found.

The average daily production of these small fish is 900 grams per square metre—over eight times the figure for the best beef pasture production in the world. The reason for the fantastic productivity is not only the continual replenishment of nutrients from below (although this is without doubt the key to the long term process), but here, in this continuously fertilised Garden of Eden, the food chain is very special. The main

Krill, the wealth of the Antarctic.

to larger and larger carnivore, to fish—there is approximately a 90 per cent loss of precious energy, and that means a huge loss of potential food for the fish population.

Long before man found and made use of the vast resource of anchovies, the local ecosystem was well aware of the potential and the shores and islands were crammed full of all manner of seabirds each with a sweet beak for anchovies. King among the anchovy eaters was the guanay cormorant which occurs in such numbers that much of the coastline is piled high with their droppings. Between 1840 and 1880 millions of tonnes of their excreta were brought to Europe under the name of guano to be used as fertilisers to bolster the agriculture that fed the expanding industrial societies of the west.

Cormorants are among the most efficient of the fishing birds and are able to locate their catch even in the murkiest of waters. There is

Anchovy fishing off Peru. There are so many you just suck them up.

plant plankton fixing the energy of the sun are colonial diatoms large enough to be seen by the fish themselves. The anchovies can thus feed directly on these plants, and are therefore herbivores only two steps in the food chain away from the sun itself. In the majority of the commercial fisheries, the food chain is much longer: it begins with small plants that are in turn eaten by small animals that in their turn are eaten by larger animals. In other words there are several steps involved before the organisms concerned are large enough to be seen and eaten by the hungry fish. At each step along the chain—herbivore to small carnivore

What a catch!

opposite
Fresh water—Antarctic ice.

good evidence that they rely, at least in part, on their acute hearing, for completely blind cormorants have been found in healthy condition. However there should be little problem for the guanay cormorants that live near the Humboldt upwelling because their fishing grounds are packed solid, and with man looking after their problems of sanitation, what more could a bird want?

The Antarctic Convergence

If you turn a globe upside down so that you can get a clear view of the Antarctic, you are looking at an ocean that is unique in a number of respects. First and foremost it encircles the world uninterrupted by any continental land mass. At its narrowest, the 'constriction' caused by Cape Horn and the Antarctic Peninsula, it is about 1000 kilometres across. The southern ice cap, unlike the northern, sits on dry land surrounded by a great ocean, parts of which freeze and thaw with the seasons. In contrast the Arctic Ocean is itself an ice cap: its centre is permanently frozen and it is ringed by the great continental land masses of North America, Greenland, and Russia. Only in the short summers do the edges of the ice melt sufficiently to reveal the narrow passages that so long tempted the early explorers in their drive both to the east and west.

Second, the Antarctic or Southern Ocean has no strict northern boundary; no land mass separates it from the Pacific, Atlantic, and Indian Oceans to the north. Nevertheless a boundary does exist that may be clearly drawn on a map. This boundary is best called the Antarctic Convergence. Here cold waters flowing north meet warm waters flowing south with a resultant change in both temperature and salinity. It is of course not abrupt but it is undoubtedly a real phenomenon, and the best way to assess its position is neither with a thermometer nor salinometer but with a simple plankton net.

The end of an era: a disused whaling station at Prince Olaf Harbour, South Georgia, Falkland Islands.

Euphausia superba or krill—the food of the baleen whales—is the signpost of the Southern Ocean *par excellence*, and as soon as the convergence is crossed, this reddish shrimp-like crustacean becomes a feature of almost every haul of plankton.

Like that of the Humboldt upwelling, the Antarctic food chain is also short: sun to plant plankton, to krill, and then to the largest animal of all, the whale, once the unchallenged king of 32 million square kilometres of cold water. The immense productivity of this immense area is maintained by the continuous upwelling of colder, denser, nutrient-rich waters from the abyss.

One unusual feature of the Southern Ocean is that its open waters are rich in life while its intertidal and sub-littoral zones are almost barren, scraped clean by the pack ice. The deep water and benthic fauna are only just being studied in detail and here some unexpected facts are coming to light. Most surprising of all is the richness of the sponge fauna, which is undoubtedly correlated with the high productivity of the surface waters maintaining a constant rain of particulate matter down into the depths. The abundance and diversity of these suspension feeders does however indicate that the lowly sponges have certainly not had their day way down in the Antarctic waters.

Of all the animals supported by this cold productive sea, the ice fish are the most peculiar. The average accepted textbook of zoology states that the bodies of all good vertebrates contain the red respiratory pigment called haemoglobin, yet the blood of certain ice fish is completely devoid of red blood corpuscles. This peculiar anaemia is partly compensated for by a larger than normal blood volume in relation to their size, but the remarkable fact is that the oxygen transport around their quite large bodies (they can be up to one metre long) is limited to that which can be dissolved in the plasma, and this would appear both physically and physiologically impossible.

The only explanation lies in their habits and habitat. From their design it would seem safe to assume that ice fish are sluggish creatures that lie in wait, camouflaged on the bottom, their large mouths ready to seize any passing prey. Although they have as yet never been seen as it were 'in the wild', it is known that they also inhabit the deep waters off the continental shelf and in fjords, where the water temperature varies only a few degrees either side of zero. Thus the ice fish have evolved to overcome the consequences of van't Hoff's factor, which is a basic law of physical chemistry stating that the rate of all chemical reactions roughly doubles with every ten degree rise in temperature.

An ice fish—a real cool customer.

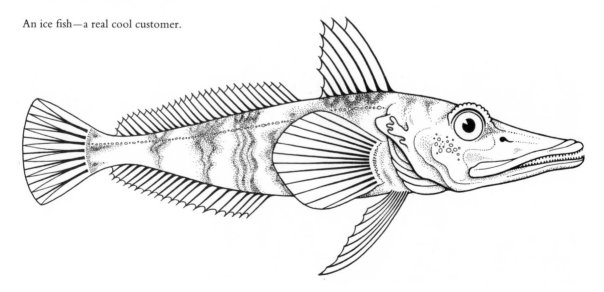

257

Because it lives at these rock bottom temperatures, its rate of metabolism and hence of oxygen consumption are also rock bottom. However if the temperature fell much lower not only would its sea water habitat freeze, but so would its unique colourless blood.

Apart from its cold, cold habitat and strange blood, the ice fish lives up to its name for it actually inhabits the wedge of water below the ice shelves that jut out from the Antarctic land mass. It is known they reside in this unlikely place because they are found 'fresh frozen' on the surface ice. The reason is that due to their bravado, or perhaps masochism, they can become trapped and frozen at the bottom of the ice shelves. Gradually over the years they are lifted up with the ice mass as new ice is packed in below, until by wind erosion they eventually pop out on the top in almost as good a condition as the day they were frozen in, which may be as much as one thousand years previously.

The coldness of the water affects not only the oxygen requirements of the ice fish but also slows the whole pace of life of Antarctic organisms. For this reason they enjoy much greater life spans than their near relatives in warmer water, and hence at any one time there are many more generations of each type of animal alive. In contrast, the diversity in the warmer and especially the tropical waters is much greater. Therefore tropical waters are characterised by limited numbers of many different species of animal, while Arctic and Antarctic waters appear to have many individuals but of a limited number of species.

Despite the open northern boundary of the Southern Ocean many of the animals that live both in it and on its shores are endemic. This is equally true for the fish, the mammals, and even the birds. There are however some animals not limited in this way. They endeavour to obtain all the benefits this world has to offer, and none more so than the Arctic tern. It summers in the high Arctic and migrates south to summer all over again in the Antarctic, feeding on the wealth of krill there. Of all the products of evolution this lucky bird must enjoy more hours of sunlight than any other living creature.

The coral industry

North of the Antarctic Convergence three great oceans, the Atlantic, Pacific, and Indian, stretch their warm inviting waters northwards across the equator and wherever they are sufficiently shallow, with temperatures never falling below 15 °C, coral animals will be at work building their massive but intricate reefs.

The coral reefs have probably caused more

The zebra fish *Abudefduf* above a grove of stags-horn coral.

opposite
A hawk fish in a garden of branched coral.

wonderment and interest than any other phenomenon of the oceans. Quite what it is that evokes this feeling in both naturalist and layman alike is difficult to say. Perhaps it is the warm clear water, the breathtaking beauty and diversity of life, the constant threat of sharks lurking in the blue, or perhaps it is the audacity of the whole operation: civil engineering, breakwaters, port and harbour installations, instant maritime real estate, all on a scale never even contemplated by man. The Great Barrier Reef alone protects the coast of Australia for about 2000 kilometres and the many reefs and atolls both large and small that rise virtually sheer thousands of metres up from the deeps are entirely due to the industry of the coral organism.

Once again, it was Charles Darwin who gave us an all-embracing theory of reef formation, explained the key role of the coral animals, and linked the facts together. For example, corals

Reef types and their possible mode of formation.

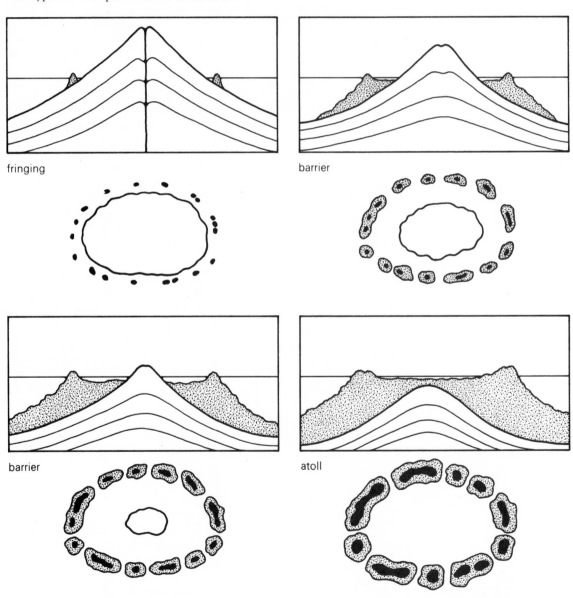

fringing

barrier

barrier

atoll

can neither live above the surface, nor in water that is too deep, and in many places where reefs are in evidence the sea floor is subsiding. He hypothesised that if the rate of rock growth kept pace with the subsidence then an island would at first possess a fringing reef, then in time a barrier reef standing further off from the shore, and finally the island itself would disappear leaving only an atoll, a ring-shaped coral reef enclosing a lagoon. In the same way, a rock platform such as a seamount lurking below the waves could be transformed into an atoll, its surface just raised above high water mark. Reef formation is of course possible in areas where the land mass is not subsiding, but in this case growth can only take place outwards to build a wider and wider reef.

Another important feature of coral ecology is that just as they are intolerant of water that is too cold, so too are they intolerant of water that is too hot. The maximum temperature in which they can survive is 39.5 °C, although some appear to be much more sensitive than others. This is one reason why coral growth is often very poor in the shallow water behind the reef ridge, where in the heat of the day the temperature may soar well above the lethal limits. In consequence the reef grows outwards leaving an area of shallow water, the reef flat, between the ridge and the contemporary shore, be it coral or bedrock. The turbulent water of the reef front, where wave and surf action stirs in plenty of oxygen, would also appear to be a factor in growth.

The Chagos Bank in the Indian Ocean gives some idea of the magnitude of the whole operation of atoll formation. Almost in the centre of this desert ocean is an area of shallow water covering some 13,500 square kilometres with scattered surface reefs around its periphery. The depth of coral rock beneath these tiny spots of dry land is probably between one and two kilometres. Whether the Chagos Bank represents a gigantic atoll in which subsidence has in recent times outstripped rock growth, or something much more complex, is not known. However only about 1000 kilometres to the north in the Maldives is Suvadiva atoll which measures 70 by 53 kilometres and ranks among the world's largest undisputed atolls. Stretching between Chagos and beyond Suvadiva is a ridge of volcanic rock that rises about four kilometres sheer up from the seabed; each atoll thus represents nearly one kilometre of white coral icing on a basalt rock cake.

A biologist diving for the first time in the environs of a living reef is in for a shock, for however well he has prepared himself, the diversity of the animal life is breathtaking. I can well remember on my first coral dive coming to the end of my air supply long before I had even finished looking at the coral head immediately under the boat.

Yet within this jumble of life there is a certain unity of arrangement, a series of patterns that in time begin to make sense. First there are those based on simple relationships between organisms: certain fish feeding only on certain plants; the territories held and jealously guarded by nesting fish; castles in the rock each occupied by a mouthful of needle sharp teeth joined to a moray eel; the fish cleaning stations where all is ordered peace; and the burrows of garden eels as regularly spaced as the trees in a well-kept orchard.

Second, there are patterns dependent on time and tide. Certain corals prefer to feed at night while others only expand their polyps during the full light of day, and some fish behave likewise or only emerge to feed when the interplay of tide and current is just right. There are patterns of destruction like those left by the large sun star *Acanthaster* as it browses its way on living coral, and the marks left by the 'beaks' of hard-headed coral fish that graze the coral rock and back away with rock flour streaming from their gills. Likewise there are patterns of regeneration as toppled coral heads begin their regrowth, the new sections orientated at right angles to the force of gravity.

The most remarkable pattern, however, is in the structure of the reef front itself. In shallow water the predominant form of the corals is massive heads, each of varied shape, some angular and battlemented, others rounded, but all able to withstand the continuous pull and

surge of the waves. With depth the massive heads give way to more delicate branching colonies that form an ordered array of off-shoots from which protrude the long coloured feeding polyps. Here there is still colour, for the light has not yet lost its full spectrum of wavelengths. Deeper again, the branching gardens give way to the rounded masses of the brain corals, each of which is delicately marked with a sinuous, repeating pattern across its surface. From this zone down the scene is washed in blue-grey light and the contoured brains give way to more flattened colonies that encrust the reef like resolidified candle wax. Here and there in this otherwise monochromatic world the scene is brightened by shapes that glow brick red or salmon pink—these are large meandroid corals which produce their own phosphorescent show of colour.

About 40 metres below low water mark the scene changes once again, for here the bulky forms are replaced by thin plate-like colonies which stand out from the reef front at right angles to the incident light. At around the 50 metre mark the reef-forming corals disappear altogether and their place is taken by communities of animals that cannot form reef rock. Here delicate yet enormous gorgonians looking like gigantic fans made of stiff lace hold themselves up to intercept or avoid the current, depending upon its strength. These grow up from a turf of silica sponges with massed arrays of soft corals and dead men's fingers, through which large sea anemones meander their ponderous way.

The groves of gorgonians soon give way to isolated patches of sea whips and solitary sponges that can grow to about the size and shape of an Ali Baba type store jar, not quite large enough to hide a man, but giving ample room for the cloud of fish that hover like butterflies around its top to disappear into at the approach of a clumsy diver. They are the last refuge, the last hiding place, for at about 55 metres the reef

Dendrophylla in deep water. This coral does not form reef rock.

comes to an end. From then on, white sand and rock debris stretch mistily downwards. This is a world of psychedelic monochrome where the slope of the sea floor imposes itself on every landscape to produce a never ending, always curving horizon that calls the diver on. But here is the place for him to turn back, for it is the edge of a desert, silent and apparently lifeless except for the slowly moving forms of marauding shark. Behind and above is the sardine-can world of the reef where so many forms of life are packed in close array. This boundary is one of the great mysteries of the sea.

The average productivity of the open Indian Ocean is less than one gram per square metre per day, a figure held low by lack of nutrients. In contrast the reef can net as much as 35 grams, with certain parts like the turtle grass meadows having a sustained level of 25 grams—production indeed.

Part of the energy is fixed by the microscopic plants living in symbiosis with the reef-forming corals, which without their brown-green partners in photosynthesis could not form their rock-like skeleton. The best proof of this fact comes from the reef front zonation already described. Only in the shallow, well-lit water do the massive branching forms predominate the underwater scene, for here, where there is more than enough light for all engineering needs, it does not matter if branch shades branch. The deeper dwelling brain corals, though massive, have a relatively smooth surface, thus avoiding the problems of self-shading, and the same is true of the encrusting forms. Further down the water column the thin plate-like colonies are each orientated to intercept the maximum possible light. Finally, at around 50 metres there is insufficient light to support reef growth and the non-reef formers take over, feeding in part on the 'crumbs' falling from the rich, productive table above.

Further proof may be found in the reef channels that carry water between the open sea and the atoll lagoon. In such channels there is no zonation: all the various forms, branching, brain, encrusting, plate, gorgonian, and sponge, are found in glorious jostling array. This is because the channel water is constantly on the move, stirred by the ebb and flow of tidal motion, hence the water is murky and full of particulate matter carried in suspension between the productive waters of both reef front and lagoon. As there is no clear-cut zonation of the light environment in the channels, and as there is a surfeit of food, competition between the various types of coral is at a minimum and the strict zonation of the reef front breaks down, again illustrating that wherever there is potential, life will use it to the full.

An atoll is a structured living system, an ecosystem, a community, each member of which has a specific function, a job to do, and it is always a job well done for the penalty is replacement by a fitter organism. The atoll itself is thus as much a part of the evolutionary process as any one of its community members. It is a productive, stable system in a desert sea.

What is the source of this productivity? Where do all the nutrients come from? Sitting at the bottom of the reef among the sea whips it is very easy to imagine a gentle current sweeping ever upwards from the deeper water. However, in many cases this is an illusion and there is no evidence to indicate that each atoll or reef is the centre of an upwelling of nutrient-rich water. The explanation is not that simple.

Imagine a platform just under the sea, lapped by nutrient-poor water. The process of reef construction can only begin very slowly because the living system is only able to draw on the small available reserves. In time, as the complex of life increases, many different food chains develop as organisms find shelter in the branching corals. Each year more and more nutrients are locked up in the cycle of the new complex web of life. Even the large pelagic predators come in from patrol at sea to lie up in the quiet water. As soon as there is any freeboard, any scrap of semi-dry land in the vastness of the ocean, it becomes a focal point for the migrating birds, and soon the message 'safe landing and free fishing' gets round the avian world, bringing more permanent residents. With them and on the surface currents come the seeds and other propagules of land plants to clothe the rock platform in a mantle of productive green. Nutrients long ago locked up in the reef rock are brought back into the cycle of the developing land system. The resident birds, such as the great frigate, the booby, and the noisy tern, radiate out to feed at sea, returning new nutrients to enrich the land and its surrounding waters.

Everything in the atoll coral garden will be lovely as long as the total system can keep a tight hold on those all-important nutrients, recycling them to be used again and again through many different pathways. Over an immense period of time the system gradually comes into equilibrium with the available supply, those lost to the abyss being replaced by additions from the surrounding sea. A coral reef is a climax system, in balance with its environment, susceptible to change, but buffered against change by its own complexity.

The dry land counterpart of an active atoll is a forest, every part of which is of equal importance to the survival of the productive whole. It is obvious that if the trees are removed the forest is no more, but in the same way if the ground flora is removed this can have an effect on the canopy of trees. Even slight changes, like the removal of the normal populations of mice and voles that feed on the forest floor, could lead to a chain reaction, eventually destroying the forest. In the absence of the natural herbivores the ground flora increases in lushness and competes with the tree seedlings, thus terminating regeneration. In the same way, a new member of the ground flora could create similar havoc by changing the environment in such a way that the canopy tree seedlings could not germinate. One infamous case was in certain forests in Bavaria where the removal of the leaf litter by the local farmers (for use as animal bedding) changed the whole nutrient economy and thus the whole ecology of the forest. A climax system is thus in balance with the total environment—the balance is dependent on and controlled by the efficient recycling of all available nutrients. It is a balance

A mini atoll, one of the three brothers on the Chagos Bank.

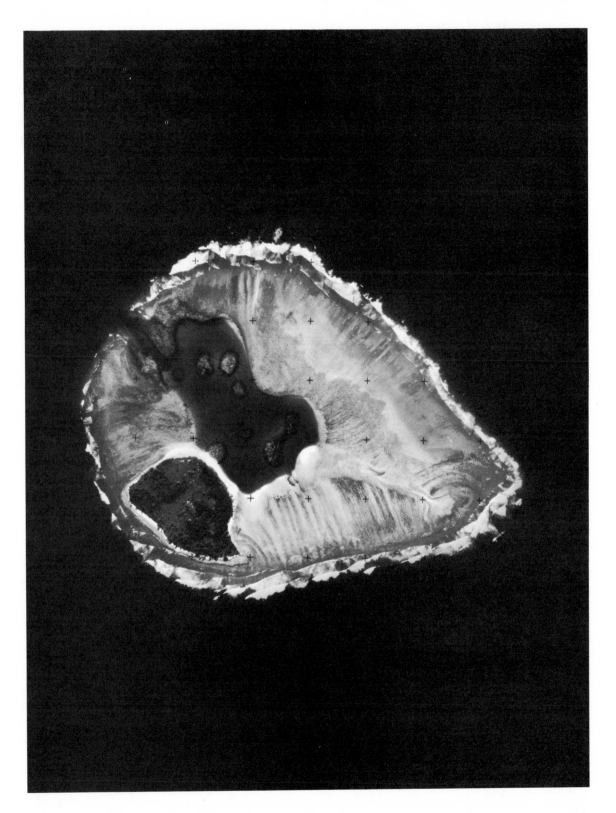

265

Fish attacking a crown-of-thorns.

evolved through time and, as the farmers discovered, one that can easily be disturbed.

If man crops a forest he destroys its evolved structure, and the same goes for all climax systems including coral reefs. Recent work has shown that in some reef areas of the world the echinoderm called the great sun star or crown-of-thorns has got out of hand. They feed on the living polyps and are therefore eating into the very resource that provides their coral home. Their numbers are soaring out of control, and the reason for this population explosion appears to be partly due to the removal from the reefs by souvenir hunters of large molluscs called tritons. One of the natural jobs of the triton is to feed on the crown-of-thorns thereby controlling its population. Similarly, a number of fish are known to eat them, so overfishing could also be operative in the demise of the reef builders.

With modern ease of world transport there are few reef areas free from the influences of man. Because it is difficult to assess the full consequences of his presence, it is very easy for a naturalist, especially one brought up near well-fished reefs, to pooh-pooh the ideas of the conservationist. My first dive on a reef that had

been free from the influence of man for over forty years made me realise just how bad things really are on most of the accessible reefs.

On this particular one, off the Egmont Isles that lie on the edge of the Chagos Bank in the Indian Ocean, not only the diversity but the quantity of fish was quite astounding. Territories were packed close together like some warring jigsaw puzzle; the queues at the fish cleaning stations stretched into the distance; and everywhere ordered masses of uniformed fish schooled their way around the coral heads. Most exciting of all, and most indicative of the unspoilt conditions, were the giant loners, the battle-scarred grand old men and women of the fish world: caranx, grouper, surgeon fish, and of course shark. As they finned their solitary way around their coral haunts they sent clouds of their smaller brethren scudding for shelter. Here was the constant war-peace, love-hate interdependence pact of the true climax system. As a visitor from another world I became a correspondent of that war but, most important, I maintained the pact of peace.

Estuaries

Of the 940 cubic kilometres of ocean water that evaporates into the atmosphere each day, 860 falls directly back into the sea (see diagram on page 282). The other 80 is all that there is left to charge the reservoirs of this dry earth, and even much of that evaporates as it recycles through the living landscapes. The rest of this sweet water moves its inexorable way back towards the sea, where it is lost, polluted by admixture with the salt water.

As a river makes its way down from its upland catchment area to the sea, it undergoes a continuous process of change. In its first flush of turbulent youth the small rivulet tumbles headlong over waterfalls, collecting oxygen and eroding and dissolving minerals from the stream bed. Hence there is only a meagre amount of aquatic life for here the key to survival is the ability to hang on tight. Later, in the serenity of meandering maturity, with its waters en-

An estuary plus man: an aerial view of the Adur Estuary, Shoreham-by-Sea, Sussex.

riched with minerals and organic matter, the river is able to support a complex web of life that adds more and more particulate matter to be carried away in suspension.

At sea level the river dies its own natural death, as with each tide a wedge of denser salt water pushes upstream beneath the river's flow. Mixing of the two bodies of water takes place at the interface of the two types, producing great masses of opalescent water that shimmer like clouds in the sun before they disappear, lost in total mix.

The estuary of any river is a zone of change, ordered in relation to the turn of the tide. It separates two very different living worlds: the sea, which nurtured the first steps of evolution, and the fresh water with its own very special diversity of life. Some animals like salmon, sea

A salmon putcheon weir on the Severn Estuary, England. The fish are caught in the wicker funnels which are invisible at high tide.

267

trout, and eels can run the gauntlet of this environmental change and can reap the benefits of both. As they migrate, their life processes have to alter to accommodate the difference in salinity for, according to the laws of osmosis, in the sea they will tend to lose water to the more concentrated milieu, while in the rivers and lakes they must guard against continuous water uptake. Just what it is that calls the salmon back to the river of its birth to lay the spawn of the next generation is not known. Neither are its mechanisms of submarine navigation understood, but the urge and the methods are there, allowing the continued existence of these fish. Each time they migrate down river to the bounty of the seas they retrace one of the paths of evolution of the animals with backbones.

There are some animals and plants that make their permanent homes in this zone of change and therefore have to adjust the balance of their body fluids at least twice in every twenty-four hour cycle. Unfortunately for them the normal pattern of salinity change is much more complex and less predictable than the regular tidal rhythm. The weather can affect both the river flow and tidal flux to produce an almost unlimited variety of change in the salt concentra-

tion, all of which must be tolerated by the permanent residents.

One feature of this harsh environment that gives the key to their existence is the abundant and continuous source of food, particulate matter carried in suspension and falling as detritus onto the bottom of the estuaries at slack water. This is the realm of the suspension and detritus feeders, and the estuarine muds, uninviting as these may appear to us, are often rich in life and incredibly productive. Hidden there are many thousands of molluscs, each with a siphon that may be short or very long but that just protrudes above the surface, working like an animated vacuum cleaner hoovering up the goodies that fall onto the mud. Their actual numbers can be astounding—500 large cockles in every square metre is not unusual. Indeed, their smaller brethren like *Macoma* can pack in ten times that population, while 20,000 tiny snails creeping about on every square metre of the surface cause neither traffic congestion nor food supply problems. Thus these muds are characterised by many individuals living it up on the richness of the suspended organic matter. However, because of the problems of salinity change there are a very few different species to be found.

Glaucous gulls on an estuarine sand bar.

Dunlins and knots fly above an estuary.

The bivalve molluscs are well adapted to the ups and downs of estuarine life, for if the environmental conditions become too bad they can simply shut up shop and lie doggo until the osmotic danger has passed. Those with long siphons like the clam are especially well suited to life in the thick, anaerobic estuarine muds, for their siphons act like snorkels, bringing down both the food and oxygen necessary for their life processes. So good are they at the oxygen transport business that very often they can be found surrounded by a halo of khaki-green oxygen-rich mud, their own personal oasis in a sea of anoxia.

The molluscan fauna of the estuaries is without doubt, and for good reason, poor in diversity of species. In contrast, the birds that come to feast on their superabundance show no such limit-ations. The shape of their bills gives the clue to the feeding habits and to their source of food: spoonbills for spooning up the food, long bills for foraging for the deepest denizens of the mud, the intermediate and shorter forms finding plenty in the upper layers, while the shortest and strongest beaks are ideal for prising open the toughest shells. Perhaps most peculiar of all is the delicate upward curved bill of the avocet, which he uses to scythe back and forth through the water to feed directly on the organisms in suspension.

In the estuaries there is not only the normal struggle for life, there is also a struggle for ownership, the sea doing its best to take over the land, and the land in turn advancing on the sea. In the temperate regions vast salt marshes spread their green fingers seawards binding and

A temperate salt marsh dominated by small shrubs.

stabilising the rich silts and in time raising them above the reach of all but the highest tides. The main silt binders in the temperate and cooler zones are quite small plants, herbs and grasses each adapted to tolerate the changes in salinity and especially the high concentrations of salt that occur when the tide is out on a hot, dry, windy day.

As with most things the salt marshes of the tropics are on a much grander scale. Here the task of building nature's own sea defences is taken over by the mangroves that may grow into trees well over 30 metres high. However, large trees with deeply penetrating roots growing in airless mud have the additional problem of keeping their roots supplied with oxygen.

A tropical salt marsh or mangrove swamp showing the arched stilt roots and spire-like breathing roots of pneumatophores exposed at low water.

opposite
Two mangrove seedlings off to a good start.

opposite
A mudskipper among the mangrove
roots.

The devil or fiddler crab *Macropipus
puber*.

The mangroves have overcome this by producing specialised aerial and breathing roots most of which are above mud level. They have also solved the problem of too much salt in two very different ways. In some the membranes of the root cells keep the salt out of the plant altogether. In others the salt taken up passes through the plant body to the leaves where it is actively secreted out onto the surface in the form of minute crystals. This can be a very good source of salt for travellers—a quick lick at a few leaves of the right type of mangrove followed by a gulp of fresh water, and an overheated tropical explorer's salt balance is quickly restored. But whatever mechanism the mangrove uses, whether exclusion or secretion. it requires energy, which is one reason why only a very few plants have adapted to this extravagant form of life. So the salt marshes, like the estuarine muds, are characterised by dense populations of only a few species.

One other problem faced by the mangroves is seed dispersal. A heavy seed falling from the top of a tree would sink right down into the oxygen-deficient mud where its demise would be imminent. Some mangroves have found the answer in early development of the seedling while still on the parent plant. It produces a stout, torpedo-shaped root on top of which the seedling sits. When shed, the torpedo falls with great accuracy to stick in the mud, and the seedling is held up safe in the oxygen-rich air where it can complete its development, putting out new roots that in time supply the mud with oxygen carried down from above.

Large trees produce many leaves and when they fall and decay they add their own abundance to the rich supply of organic matter that in turn supports an abundance of other life. This is the home of the most arboreal of all fish, the mud skipper, which can spend long periods of time high and dry, clambering about among the mangrove roots. Here too the fiddler crab makes his weird tic tac signs with his outside claw. It is certainly most extraordinary to watch. The whole crab colony can be busy feeding, their small claws and mouth parts working ten to the dozen, when all of a sudden, almost as if one of them gave an order, many of the crabs will stop, stand on tiptoe on their walking legs, and wave their outsize claws in the air. Once apparently satisfied, they return to the more serious job of eating. Just what the signal means is not known, although it is thought to be a ritualised 'this is my territory' signal. However it does not look menacing; on the contrary, it appears to be the most affable of gestures. Who knows, it is probably merely an indication that the tropical mudflat is an ideal place for a fiddler crab to live, with plenty of food for all the happy, gesticulating population.

273

Farm and conserve

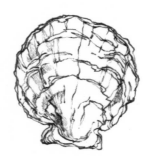

It is impossible even to hazard a non-academic guess at the date when man first took to the oceans in search of food. With each tidal cycle the retreating sea must have lured him on with rich finds of shellfish down to low water mark, and then further out as he developed his ability to swim and dive.

Apart from duration and depth, and we know that the modern breath-hold divers of Indonesia are capable of excursions to depths in excess of 100 metres, his main limitation underwater was poor visibility. Eventually this was overcome when the invention of clear glass made first the window bucket and then the diving helmet and face mask possible. Once good visibility was the order of the dive, the underwater harpoon and eventually the compressed air gun became the disorder of the day. At the same time, fishing from the surface went ahead in leaps and bounds widening man's horizons both spatially and gastronomically. The expanding use of fish traps, nets, lures, hooks, lines, rods, reels, and dynamite increased the efficiency of man, hunter of the seven seas.

The modern multi-million dollar fishing fleets with their aerial reconnaissance, plethora of electronic devices, suction pump fishing, freezer factory ships, etc, etc, etc, are simply twentieth century extensions of the technology of the hunt, which is on such a massive scale that in 1970 alone the world catch was 70.4 million tonnes. There is little doubt that this technology will be refined even further, bringing with it the warning of potential over-exploitation. Yet it has been calculated by the United Nations Food and Agricultural Organisation that the world fish stocks could provide almost twice this figure with no foreseeable long-term problems.

An efficient hunter of the sea, a modern stern trawler.

Trappings of the lobster hunt.

However accurate their predictions, it would appear that at least as far as world fisheries potential is concerned, time is on man's side. Nevertheless, there is a natural limit and that limit can only be exceeded by turning from natural to 'unnatural' means, from hunting to husbandry. While there is still time man must learn to farm the sea.

There is already a long and honourable history of marine farming, experience and success on which future developments can be based. The earliest records of mariculture relate to the cultivation of oysters in Japan well over 2000 years ago. It seems that throughout history oysters have been great favourites, and only in recent times, when over-exploitation and pollution have decimated the natural estuarine oyster beds, have they become the expensive luxury they are today. In the past, both in Britain and on the continent of Europe, oysters were plentiful and cheap and a good source of protein for the common man. But early in the eighteenth century laws had to be passed in France to save the declining over-fished oyster beds by restocking them.

Oysters are eminently cultivatable, for a single adult can produce as many as one hundred million eggs at one laying. As the fishermen imagined that the oyster spat out the eggs, the egg mass has come to be referred to as the 'spat'. Each egg develops into a veliger larva that swims in the plankton for two to three weeks before settling down on a suitable substratum. Beds are prepared to receive the 'spat' by laying down

cleaned oyster shells or porcelain, called 'cultch' in the trade, and any lurking oyster predators are removed by hand prior to fencing. The oyster field is then ready, but the harvest will not be ready for a few years because even in the best estuarine conditions it takes a long time for an oyster to grow to full size.

This method of husbandry has its limitations, the main one being that the beds only benefit from the food that falls to the bottom. In recent years the traditional oyster beds have been replaced by suspension cultures in which the cultch is hung from ropes attached to floating frames or to stakes driven into the bottom. The total food supply in the full depth of the moving estuarine water is thus brought into play and the oysters are suspended safe above their bottom-dwelling enemies. Using such techniques it is possible to obtain production levels of 6400 tonnes of oyster meat per square kilometre (6400 grams per square metre) of farm.

When it comes to hanging around in suspension cultures the high class oyster cannot compete with the very common mussel. Mussel cultivation pioneered in the vicinity of Vigo in northern Spain can net a fantastic 27,000 tonnes of mussel meat for each square kilometre of float.

The success of the inshore cultivation of seaweeds is referred to on page 54. Laver bread, once a small part of a subsistence diet of peasants in various parts of the world, is now big industry in Japan and a significant entry on the menus of Chinese restaurants the world over. Success with other seaweed types is already heralded, one of the more interesting cases being that in Haiti of *Laminaria japonica*. Like most of its near relatives, this large brown seaweed, a highly sought after delicacy, is a plant of cool temperate seas and will not live in warm waters. Recently it has been found growing in water well above its normal temperature range, but in every case it

opposite
A natural bed of natural oysters:
Ostrea edulis.

An oyster farm: a tray of Pacific oysters *Crassostrea gigas* basking in warm water, the waste heat from a power station.

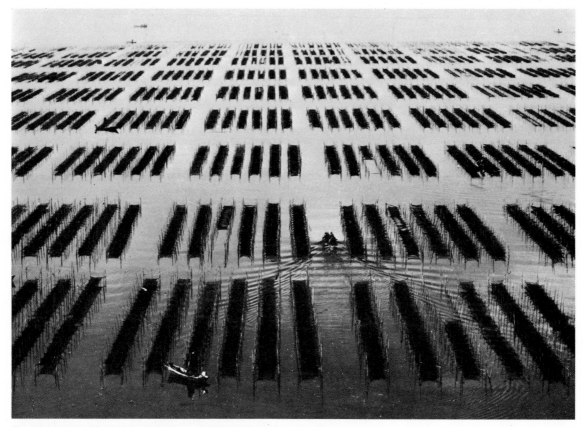

Farming seaweed in Japan. Each of the 'fields' consists of stakes driven into the silt on which the nori is grown.

is in water polluted by sewage nutrients. This has opened up the possibilities of extending the range of Haiti cultivation with cultures enriched with fertiliser.

In fish farming, the record of success has been first class although it has only just started on anything like a massive scale. *Chanos chanos*, the milkfish that has been farmed in Java since the fifteenth century, has recently come into wide use, with some startling results. The fry are collected in vast numbers out at sea and are placed safe from their predators in shallow fish ponds filled with estuarine water. In the Philippines such farms using unenriched seawater produce 31 tonnes of fish per square kilometre. In Taiwan, using fertilisers added to the fish ponds, the figure stands at 206, while in Indonesia the addition of human sewage nutrients boosts the production to an incredible

508 tonnes per square kilometre. These figures must be compared with 7 tonnes which is the natural level of open ocean production.

There are problems, especially in the sewage enriched cultures where human diseases and parasites could multiply and be transmitted through the fish products. Nevertheless the potential is there and methods to overcome most of the difficulties have already been developed. The United Nations has calculated that in Southeast Asia alone there is at least 5500 square kilometres of shallow sea that could be turned over to milkfish production. If it were farmed using the Taiwan enrichment methods, the milkfish farms could produce more than the world's contemporary total fish catch each year.

Another main difficulty is the capital outlay for setting up and especially for the civil engineering needs of such schemes. Wherever direct fertilisation of the sea has been attempted problems have arisen. First and foremost, fertilisers are expensive and it would be a foolish man who threw his money into the sea, especially if there was any chance of his nice fat fish swimming off into someone else's net. For this reason most of the experiments have been carried out in fjord-like bays whose entrances can be closed by bubble, if not by concrete, barriers. Then the fun really begins. The fertiliser simply disappears, being sopped up by the bottom muds and the adventitious members of the natural benthic ecosystem. Then, when all the natural nutrient reservoirs are satiated, and the plant plankton, followed by the zooplankton and

Seaweed drying in the sun down on a Japanese farm.

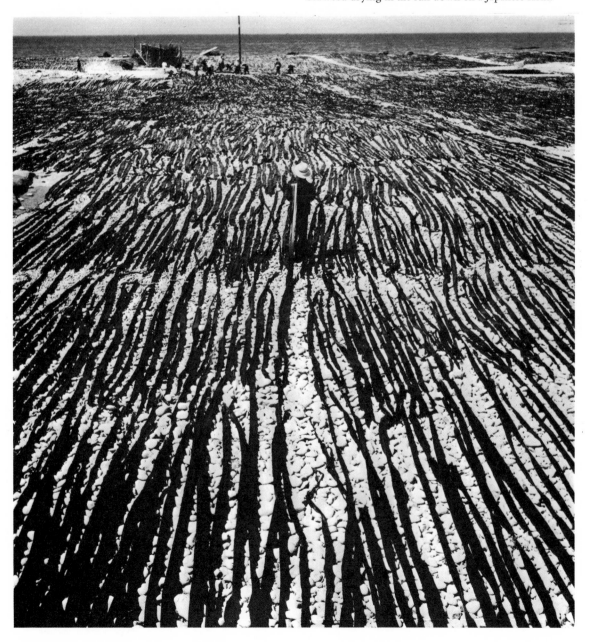

finally the fish begin to show an increase, the farmer is in business. One problem is that the whole local ecosystem soon gets to know and in come all the fish-eating birds ready to gorge themselves.

It is for these reasons that time and time again brave new fish farmers have turned from the open sea, invested their capital in civil engineering works or in floating cages, and gone in for rearing big profit, expensive-type fish. There is 'turbot au gratin', 'lobster thermidor', and 'real' turtle soup doing well in marine farms in many parts of the world, but the day of the factory-farmed cod and chips (French fries) still appears to be a long way away. There is however no getting away from the fact that the breakthrough will come and farming the sea on a massive scale will become a practical reality. With the world's problem of malnutrition, it must.

One of the most fascinating schemes envisaged to date stems from the active mind of Oswald Roels and his colleagues of the Lamont-Doherty Geological Observatory of Columbia University. They hit upon the idea of tapping the natural supplies of cold nutrient rich water down below the 1000 metre mark off the island of St Croix in the Virgin group. A pump brought up the water through approximately two kilometres of wide-bore plastic tubing, and fed it into tanks in which the local plankton grew twenty-seven times as fast as they did in the natural state—an ideal basis for a fish farm.

So it was that a most fantastic idea of a package deal farm of the future evolved. Take one coral atoll, with not too deep a lagoon and conveniently narrow reef channels; lift nutrient water from below 1000 metres by means of a pump; and let it, in part, siphon over into the lagoon. With nutrients galore and on the cheap, up goes the plankton production—just what the farmer ordered. Furthermore, already there are spin offs. Cold water at 4 °c lifted up to tropical heat could be put through a heat exchanger and the temperature difference used to air condition buildings, run a refrigeration plant, and condense fresh water from the tangy island air. The possibilities are endless: luxury hotels, fresh water for drinking, irrigation or hydroponic farming, an abundant supply of fresh fish, and freezing capacity to store surplus for export. A fish farm, a shark farm, even a whale farm, all could be an added tourist attraction, thus making full use of the marine environment. And this is not science fiction; it is science fact.

It was at the Oceans 2000 Conference held in London in 1973 that the divers of the world came to listen to, among others, Arthur C. Clarke. In his own inimitable way he painted a verbal picture of our own inner space odyssey: mineral extraction based on seaweeds and invertebrates bred to concentrate elements from the infinitesimally small quantities found dissolved in sea water; the transshipment of icebergs behind nuclear tugs to places where water is required for irrigation; massive utilisation of the temperature difference between the abyss and the euphotic zone to generate electricity; underwater turbines situated in the great water currents that stir the ocean deeps; the ranching of whales not for meat but for the tonne of milk they produce each day; and man-made upwellings bringing fertility to the desert oceans. All this and more he put over in his easy style, raising many a laugh with his superb jokes. He was not joking however when he said: 'We all know which is the most terrifying, most destructive creature in the sea. He is right here in this hall. That is why we at this congress have a great responsibility. The question this generation must answer is, "can we exploit the sea without destroying it?" '

At least in one sphere of endeavour we are going in the right direction, for the future of fish farming looks good. If the potential milkfish farms in South-east Asia could provide the world with an adequate amount of fish protein, and if all the other projected farm systems could add the variety necessary to make farmed fish as attractive as the natural product, many of the natural fish stocks of the world could be left alone to give them a chance to rehabilitate after centuries of exploitation. The only 'problem' would be that the sentiments of the famous Satchmo/Bing Crosby song 'Gone Fishin' would no longer be the same, for 'fishin' would be synonymous with 'farming'.

Sink and resource

Since the time of Adam and Eve, some 110 thousand million human beings have lived on this planet. If they each averaged a twenty-five year life span—a moderate estimate—and each produced about two and a half litres of urine per day throughout their lives, this would mean in round (or rather square) figures that 2500 cubic kilometres of human urine has been produced and got rid of, its ultimate destination being the sea. But even if the total volume of urine were still there intact, it would represent little more than the proverbial spit in the ocean, for the total volume of the seas is in excess of 1300 million cubic kilometres.

Every day 940 cubic kilometres of water evaporate from the surface of the sea and pass into the atmosphere to eventually fall as rain. The major part of it falls directly back into the sea; in fact only about 80 cubic kilometres supplies the dry land with water. Therefore the sea is not only the source of all marine life but of all life on earth.

It is interesting to analyse these figures in a little more depth. Eighty cubic kilometres represent exactly 80 million million litres of fresh water—an awful lot. A healthy man requires about three litres per day to keep his metabolism quenched, which is not really very much. But an average man living in the technological western world needs 70 times that, in other words about 200 litres a day to flush his super loos, work his automatic washing machine, and shine his horseless carriage. Add to this the following staggering statistics: 220,000 litres of water are needed to produce one tonne of steel, 60 litres to distil one litre of petrol, and so on and so forth, plus the amount of cooling water needed to make electricity, the water required for irrigation in agriculture and horticulture, etc, etc, etc, and all of a sudden clouds become about the most important feature of everyday life. A simple sum of very long division (the volume of water that falls each day as rain, divided by the present world population, which stands at around 3600 million) shows, to put it mildly, that some of us are getting far more than our fair share.

Now and in the future man must look more and more to the resources of the sea to sustain his many needs, but the problem is that the sea is not only the world's last major untapped resource, it is also the final sink for all products of erosion whether natural or man-made. Modern technology has not only taken man into the sea, it has turned him into a great geomorphological and geochemical force. Every day larger and larger quantities of material derived from the earth's crust flow out through our short term economy to end up in the sea. Some are inert, some are beneficial, but many are detrimental, and all are drastically changing the marine environment.

Evolutionary fitness is compounded of two interrelated systems: the fitness of the organism and the fitness of the environment. If the latter is changed the former must change also. The living system or ecosystem of the sea is buffered against change, the component systems handling what they have evolved to handle. Slow changes in temperature and salinity, localised rapid, even catastrophic, changes such as those caused by volcanoes, earthquakes, or *tsunamis* can all be accommodated by restocking from adjacent areas.

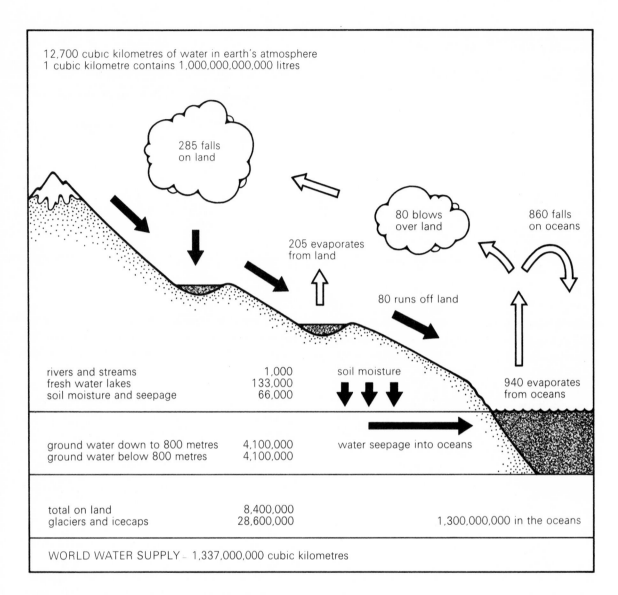

The following labels appear within the diagram:

12,700 cubic kilometres of water in earth's atmosphere
1 cubic kilometre contains 1,000,000,000,000 litres

285 falls on land

80 blows over land

860 falls on oceans

205 evaporates from land

80 runs off land

940 evaporates from oceans

rivers and streams	1,000
fresh water lakes	133,000
soil moisture and seepage	66,000

soil moisture

water seepage into oceans

| ground water down to 800 metres | 4,100,000 |
| ground water below 800 metres | 4,100,000 |

| total on land | 8,400,000 |
| glaciers and icecaps | 28,600,000 |

1,300,000,000 in the oceans

WORLD WATER SUPPLY – 1,337,000,000 cubic kilometres

The world's daily water cycle (in cubic kilometres).

With man's rapidly developing freedom of the oceans a new question must be asked and answered: can he affect the fitness of the marine environment in such a way that it becomes unfit for the continuance of the process of evolution? Can man pollute the sea?

The cold North Sea:
a cautionary tale

Man is an integral part of the process of evolution and he must not be allowed to forget it. Just as at some time in the dim and distant past the first prototype whale lumbered its way back into the sea thus escaping from competition with the burgeoning population of land-bound mammals, so too is man in the guise of *Homo aquaticus* returning to the sea, tapping its bounteous resources to feed his exploding population.

Already in his struggle for existence, he has eradicated the great auk and the largest of the sea cows from the face of evolution; in the same way he may have sealed the fate of the whale, the largest animal that has ever lived. These are acts of pure vandalism: removal of even one thing from the total marine resource may cause irreparable harm for the whole system, including man. One interpretation of the Darwinian principle of evolution gives man *carte blanche* for such acts, in the belief that the top dog will always survive. Another, which recognises the importance of the ecosystem or structured whole, casts doubt upon the very fitness of man and poses the question, will he survive?

One answer would be to ban man from the sea, to halt his progress at the edge of the oceans, thus stopping *Homo aquaticus* in his tracks. But this would be both impracticable and wrong because as part of evolution man has an equal right to a fair share of those resources. Every day he becomes more and more dependent on the life-giving sea for his very existence, and this dependence on the marine resource is bound to continue to rise. He cannot stop making use of it; all he can do is conserve it, that is, use it wisely. Over the last two or three decades a new

concept has developed in the social evolution of man, a concept that may gain him survival—the concept of conservation.

Of all the seas that have been both used and abused by man, perhaps the cold North Sea can boast the longest history of certain types of misuse. Its shores saw the birth two centuries ago of the industrial revolution, which was not only to change the lives of the human population but was to affect just about every other living organism. Being part of the continental shelf of Europe now flooded by sea water, the North Sea is relatively shallow. Less than 40,000 years ago the bulk of the shelf stood high and dry and the land masses were covered by the ice sheets of the last glaciation. The water locked up in the solid ice sheets had been lost from the sea by evaporation and this caused a reduction in sea level of about 100 metres. These effects were not limited to the cold polar seas because all the oceans form a continuous single reservoir, so glacial draw down was a world-wide phenomenon.

During this time the reefs of the coral seas must have protruded from the surface as great pillars of limestone rock eroded by the onslaught of the waves, while below the new sea level new reef was being laid down. The remains of these glacial period reefs have been found encircling atolls in the blue grey depths where contemporary coral growth is impossible due to insufficient light.

Throughout the ice age what was left of the North Sea would have been, if not completely icebound, at least subject to the formation of pack ice, and its edges would have been scraped clean by ice action. Today its 54,000 cubic

kilometres of water rank among the most productive seas of the world providing no less than 5 per cent (about thirty million tonnes) of the world's annual commercial fish catch.

The source of this productivity is complex. First and foremost, because it is a shallow sea, light penetrates throughout much of its depths, so there is no large, deep, dark abyss in which nutrients can be locked up out of recycling's way. Second, each year the river Rhine alone adds about 70 cubic kilometres of water charged with nutrients and other chemicals eroded from the landscapes of natural and industrial Europe. More important still, each year 22,000 cubic kilometres of water sweeps in from the Atlantic around the north coast of Scotland, bringing with it rich plankton, many of them denizens of the warmer waters of the south, and a new supply of nutrient salts. The residence time of this added water is some two years and during this time the planktonic ecosystem draws on the nutrient supply before it eventually passes to the benthos, which is rich in suspension and detritus feeders. Although the water body circulates and moves on, some of the nutrients, trapped in the food web, are retained for longer periods, enriching and bolstering production. The North Sea is thus a vast and renewable resource of high-grade protein.

Long before man started to make use of these stocks of fish the natural ecosystem had whistled up its fair share of gourmands. At least seventy-one different sorts of birds from all over the globe, whether migrating through or spending part of their lives there, feed on its wealth of life. Some, like the Arctic tern, make short stop-overs to refuel on their long pan-world migration flights; others overwinter, probing the mudflats of its many estuaries for their food. Twenty-nine of the species moult in its sheltered food-rich habitats, and during this significant phase in their life cycle not only are their powers of flight and insulation impaired but much energy is required to grow a new set of feathers. In addition, twenty-five species spend the most important part of their life cycle there, breeding on sea cliffs and rocky islands out of the way of the main effects of man.

Peter Evans, an ornithologist who has worked extensively on the ecology of migrating birds, has calculated the following figures which, although very approximate, are of great interest. The average dry weight of birds using the North Sea each year is 550,000 kilograms, and there are about 5 million of them. The energy required to keep this lot flying, wading, moulting, courting, nesting, brooding, breeding, and shrieking is about 9×10^{10} kilocalories, which is

opposite
Glaciers on the edge of an ice-bound sea.

Two knots of knots. These birds migrate over the North Sea.

less than 0.01 per cent of the total primary production of the North Sea. However as there is roughly a 90 per cent loss of energy at each exchange step in the food chain, and as many of the birds represent the final stage in complex chains, they are certainly taking their fair share out of the system. Detailed study of their feeding shows that they have little or no direct effect on the commercially important fish, although they must compete with them for some of the available food. For example it has been calculated that in the Wadden Sea the birds take between 10 and 20 per cent of the mud-living molluscs, and the birds eat what the fish cannot enjoy. However even this is an indirect form of competition, for the birds feed on the adult molluscs while the fish in the main feed on their planktonic larvae.

In comparison to the seventy-one different types of birds the ornithologist can enjoy around the North Sea, there are at least thirty different sorts of fish that his gastronomical counterpart can savour, with or without chips, although only eleven of these are of real commercial importance. They are best considered in two groups. Cod, coalfish, haddock, plaice, sole, and whiting are all eaten directly by man, and

Eider down by the North Sea: an eider duck and drake.

opposite
Brünnich's guillemots and common guillemots which breed on the coast of Norway.

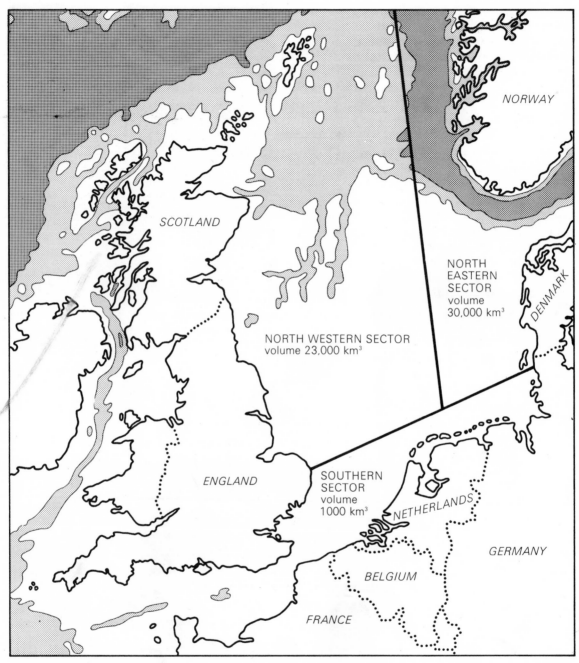

NORWAY

SCOTLAND

NORTH
EASTERN
SECTOR
volume
30,000 km³

DENMARK

NORTH WESTERN SECTOR
volume 23,000 km³

ENGLAND

SOUTHERN
SECTOR
volume
1000 km³

NETHERLANDS

GERMANY

BELGIUM

FRANCE

SEA DEPTHS

☐ <100 metres
▨ <200 metres
▦ >200 metres

Depths and sectors of the North Sea.

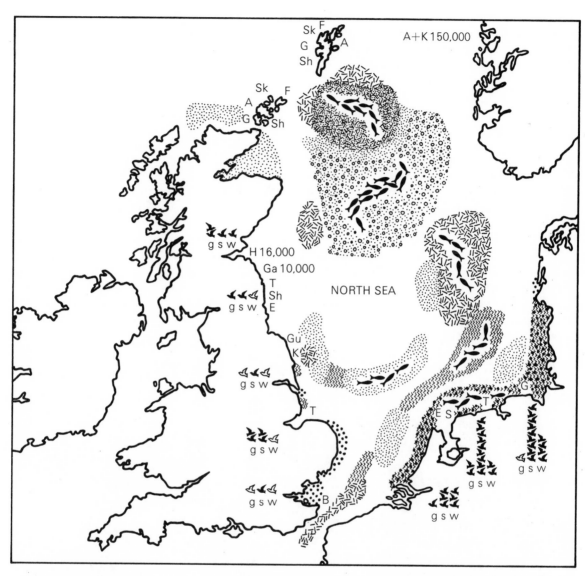

Sk F
G A
Sh

A+K 150,000

Sk F
A
G Sh

NORTH SEA

g s w
H 16,000
Ga 10,000
T
T Sh
g s w E

Gu
K

g s w
T

g s w

g s w B

G

E S T

g s w

g s w

BREEDING COLONIES

A   auk
B   black-headed gull
E   eider
F   fulmar
Ga   gannet
Gu   guillemot
G   gull
H   herring gull
K   kittiwake
Sh   shag
S   shelduck
Sk   skua
T   tern

BIRD POPULATIONS

g   gulls
s   shore birds
w   wild fowl

      <50,000

      <100,000

      <200,000 etc

AREAS OF IMPORTANCE FOR
NURSERY AND/OR SPAWNING

      cod

      herring

      haddock

      plaice

      sole

North Sea map showing the areas of importance to the
fishing industry, the bird populations, and bird breeding
colonies.

289

they are mainly demersal fish, that is, they live or feed on the bottom. The other group, herring, mackerel, and sprat (which live in mid-water), and Norway pout and sand eel (which are demersal), are not usually eaten directly but are used to make the fish oil and fish meal of animal feed. Fish protein, to animal protein, to man, (or even worse, fish protein used as a soil conditioner, to plant, to animal, to man) puts man at the end of a long food chain that has very large natural losses in potential all along the way. In a world where much of the human population lives on the brink of protein malnutrition this is crass stupidity, a definite misuse of a very valuable resource.

The population of any one type of fish depends on a complex of factors: the production potential of the volume of sea in which it lives, the regularity of the breeding cycle, the number of viable fertile eggs, and the fry's ability to successfully run the gauntlet of the ravenous food chain

to attain reproductive adulthood. The success of the hatchings determines the maximum number for recruitment into the population. After reduction by natural causes, those remaining after a certain period of growth and development are recruited into the next age or size class of fish. The important point is that enough fish must reach reproductive maturity so as to ensure that the process can continue. Once a fish is past its reproductive prime it becomes supernumerary to the generation game, although it is still an important source of food for any predator with a sufficiently large mouth.

In time, the fish population comes into quasi-balance with its total environment, the standing stock depending on the potential number of recruitments into each age or size class and the number of mortalities, whether natural or man-made. Many factors can affect the balance. For

A North Sea fishing fleet in harbour at St Monance, Scotland.

example, a bad winter or summer with reduced production of plankton can increase competition for food and hence reduce numbers. Similarly disease, often wiping out much of one age class, can produce a population gap, causing the actual standing stock to fluctuate from year to year.

Man puts himself into the picture by stepping up the number of mortalities. The perfect state of affairs would be to crop only the excess production of each year to ensure that sufficient stock remains for adequate recruitment. It is the job of the fisheries biologist to study the structure of the population and thereby predict both standing crop and sustainable yield. He can therefore assess the well-being of the fish stocks and the effects of man's predation. But it is rather like the scientifically-based fishing industry playing a game of cards with the environment—the fisheries biologist is well versed in the rules of the game and, given sufficient information, should always win, but the environment can hold some unpredictable and devastating trump cards.

Between 1950 and 1970 the tonnage of fish removed by man from the North Sea more than doubled, from a figure of about 14 million to about 30 million. The reasons for this are manifold and each species should be considered separately. In essence, however, the increased landings of the first group, the fish-and-chip-type fish, are due to increased recruitment to the population, an increased biomass of fish. In contrast, the increased catches of the second group, the fish-meal fish, are due to increased exploitation by man resulting in a reduction in the standing stock. It would be just too easy to infer that the increased population of the first group has been caused by the reduction in the numbers of the second, the pelagic fish no longer competing so strongly for the food resources of those living on the bottom. Ecology is unfortunately not that simple and the reasons are still obscure. But whatever the answer, it is a fact that in recent years the demersal fish have definitely prospered, much to the delight of the fish-and-chip trade and, one presumes, to that of the inshore birds who actually feed on their nursery grounds.

Important as the open water of the North Sea is, the role of the much shallower inshore areas in the total biology of the sea cannot be overestimated. Many of the more succulent fish, like sole, plaice, cod, haddock, and herring, have their spawning and nursery grounds close inshore. Add to this the rich populations of shrimp, crab, lobster, oyster, mussel, cockle, winkle, and whelk, each floating their myriad

Total fish landings from the North Sea.

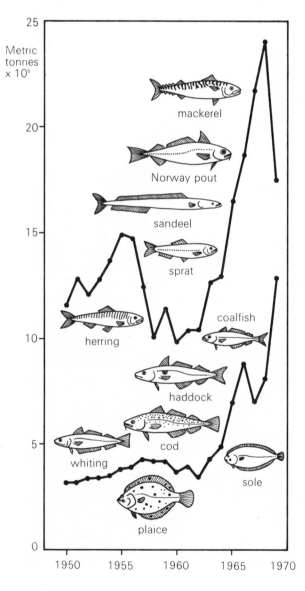

British tradition Chelsea-style—everything, well almost everything, with chips.

larvae up with the plankton, and it is obvious that this sea must be considered as an integrated living community whose members all have an important role to play in the marine resource.

At least part of the high productivity of the inshore areas is due to the proximity of the surrounding land masses which discharge through their river systems the valuable products of erosion. Here, nutrients brought from the land enrich the waters that feed the close-packed populations, and wherever bare rock gives a firm foothold, some of the most productive systems of all are found. Thick masses of wrack cover the littoral zone with a blanket, beneath which many animals find damp shelter during the periods of ebb tide. Below the low tide mark the vast forests of kelp produce ever changing patterns of shade, glades of protection for a cross-section of the animal kingdom, and

shelter for the diverse inshore web of life. Of all the living systems of the sea these are the ones that must bear the brunt of the effects of man, for this is the boundary he must cross to reach out into the sea.

What are man's effects on the resources of the sea? Today he must rank among the most active if not the most efficient predator of North Sea fish. This almost certainly alleviates some of the competition for the basic food stocks, and certain food organisms might well heave a sigh of relief as man, their saviour, appears on the scene. However, the resources of the oceans are not limitless and their populations are contained and controlled by the lack of some essential factor in their diet or by falling foul to some other carnivore.

Direct massive predation is only one way in which man affects the marine environment. His at first sight innocent but greedy use of the sea for pleasure and education can have far reaching effects. Already recreational pressures are building up as 'sport' fishermen jostle for their share of the dubious pastime, and the constant rain of tin cans and bottles falling to the bottom around the plethora of overcrowded marinas is providing new, fully detached homes for a staggering variety of life, both plant and animal, vertebrate and invertebrate. Even those who are only interested in the wildlife itself are beginning to endanger the very animals they come to see. In 1953 the Farne Islands, a small group off the north-east coast of England, were visited by a mere 2500 tourists and naturalists, all of whom go to enjoy the crowded colonies of seabirds and the bobbing heads of the Atlantic grey seal. In 1971, only eighteen years later, the number of visitors had risen to 25,000—more revenue for the maintenance of the reserve, but also more disturbance for the birds and the plant and animal life of the well-trodden paths.

One of the most fantastic spectacles of the Farne Islands can be seen during the mating season when the full breeding colony of the grey seals haul themselves out of the water to wallow in the peaty mud and raise their soleful pups. This is one of the rarest of the world's large mammals and in the past it was slaughtered mercilessly both for seal oil and for the soft pelt of the young pups. The cry of the conservationists was heeded as long ago as 1914, and the passing of a parliamentary act gave the population the necessary protection and thus helped to save this rare and beautiful animal from almost certain extinction. Since that date, safe on its Farne Island sanctuary, their numbers have steadily increased until today they stand in excess of eight thousand.

Unfortunately this population explosion is causing grave concern among the local fishermen on at least two counts. Seals certainly compete for the rich salmon fishing of the Tweed and adjacent rivers, and however much one argues that they are only taking their fair share, they do both damage to and remove the fish from the nets, which understandably is enough to make any fisherman angry. In addition they are a carrier of the cod worm and hand in hand with the seal increase has gone an increase in the infestation of the cod catch by these revolting parasites. Although it may not be cause and effect, the indications are certainly there that fisheries and a protected seal population do not mix.

Test culls have already been put into operation, a certain number of animals being removed from the population, and the results are awaited with interest both by fishermen and conservationists alike. There are signs however that the seal population is already being limited by the size of the island refuge. As their numbers have increased, more and more seals are forced to take up territories near the middle of the islands.

Relationships among North Sea resources.

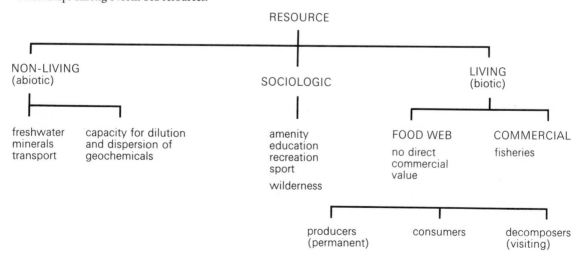

Atlantic grey seal bull.

This means that the pups have to travel further from the safety of mum to their first swim in the briny and the route is crowded with other large wallowing members of their own species. The result is a much higher pup mortality rate for the inner island dwellers and this could stabilise the population. It is of great interest that the animals that are physically or socially inferior are the ones to take up the inferior residential sites, so the strongest and more aggressive members of the population will come out on top, reproduction-wise. Population control combined with survival of the fittest thus makes for 'bigger' and 'better' Atlantic grey seals.

The lesson of the 1967 *Torrey Canyon* disaster which opened the eyes of the world to the realities of marine pollution by oil, spells out stark warnings as news from the North Sea rigs tells of new finds of oil and natural gas. Accidents will happen, and many birds will die the foulest of deaths, choked with the thick, black oil, and if this happens near one of the large breeding colonies the slaughter will be enormous.

Since *Torrey Canyon* much money and research has been invested to find methods of cleaning up the aftermath of such an accident. The lesson hard learned on the beaches of Cornwall was that, apart from the birds, the ecosystem could ride the onslaught of the floating scum of oil and given a chance could rehabilitate with surprising rapidity. The worst damage was done wherever detergents were used (all in good faith) to emulsify the oil so that it could be flushed away to mix with the sea water. The problem was that the oil-detergent emulsion was very toxic to many inshore animals and the slaughter of the benthos was even greater, though neither as visible nor emotive as that of the seabirds. The French authorities got rid of their share of the problem oil by sinking it with a blanket of, what else but, french chalk. The birds were thus saved from a

opposite
Chemical warfare—the fight is on. Detergent being sprayed on oil covered rocks at Sennen Cove in Cornwall.

An oiled gannet.

fate worse than death and the benthos was kept detergent free. However, in time the oil-chalk mixture worked its way along the sea bed, ending up on fishing nets and holiday beaches where it then had to be removed by hand—a long, costly, and messy job.

Oil will continue to be an environmental problem for as long as there is some to be found, and with the rising price of petroleum and its products, and the enormous effort being put into the discovery of new resources, it is likely to remain with us for some time to come. Fortunately, out of the black murk of *Torrey Canyon*, one good thing has emerged: no longer do the great tankers of the world flush out their oil into the sea. An ingenious method, called load-on-top, of re-flushing the tanks and re-taining the bulk of the dirty water on board, has almost eradicated this serious source of marine pollution. All the big tanker fleets now practise load-on-top and refineries have agreed to accept a certain amount of the oil-water mixture with each new load of crude. This simple code of international practice has greatly reduced the oil pollution of the high seas, but it has increased the price of oil, which must be borne by the consumer.

This raises a question of fundamental impor-tance: just how many concessions can the economy of twentieth century man afford to make when it comes to environmental care and hygiene? Indeed, in the final analysis, how many should he make? Although it is at present impossible to give an answer, it is clear that with the limited resources available for environmental problems, it is essential that what money there is be spent wisely.

If a value were to be placed on each aspect of the North Sea resource, by far the greatest would be on the power of the sea to accept, dilute, and disperse the many products of our affluent society. Its shores have cradled the latest attempt at civilisation, an attempt that over the last two centuries has become increasingly dependent on industry. It was the sump of the industrial revolution and if it has any claim to a place in the record book, then that must surely be as the sea with the longest history of pollution by an industrial society, a society that draws on the resources of the world. The immense tonnage of guano brought into Britain during the period of the industrial revolution highlights this dependence. The guano was needed as fertiliser to boost the agricultural output of the British countryside for food to feed the factory workers. Until this time the majority of the people worked on the land and formed part of a semi-natural food chain. The evolution of industrial man meant that fewer agriculturalists had to produce more and more food to sustain the hungry workers, the products of whose labours were then used to purchase food, fertilisers, and raw materials from many other countries.

The industrial revolution was just a foretaste of what was to come. In 1970 alone the east coast ports of England handled at least 184 million tonnes of exports and imports, just part of the flux of material needed to maintain the metabolism of the agro-industrial complex that forms the very English ecosystem. The by-products of the same metabolism, 450,000,000 litres of human sewage and 800,000,000 litres of trade wastes, pour into the North Sea each day, affecting the environment. These products of our 'effluent' society are best considered under four broad headings: enrichers, suspenders, cloggers, and killers. The four categories are not mutually exclusive, and it is possible for one substance to play a number of polluting roles.

The enrichers consist of plant and animal nutrients, compounds of potassium, nitrogen, and phosphorous. These are best called eutrophicants as they enrich the water, increasing its productive potential. Taking all the man-made sources of nutrients entering the North Sea into consideration, they are insignificant when compared to the enormous input from the Atlantic. Nevertheless man's additions to the nutrient budget are made in the shallow inshore waters and more often than not create hot spots of eutrophication where the localised increases in primary production are rapidly balanced by decreases in light penetration. The same is true of all other pollutants: near the infalls there will always be hot spots of change. Their size will depend on the magnitude of the infall, and their effects will rapidly diminish as the substance in question is diluted and dispersed.

The suspenders directly affect the clarity of the water by reducing light penetration, and

Lights beside the sea: the atomic power station at Dounreay, Scotland.

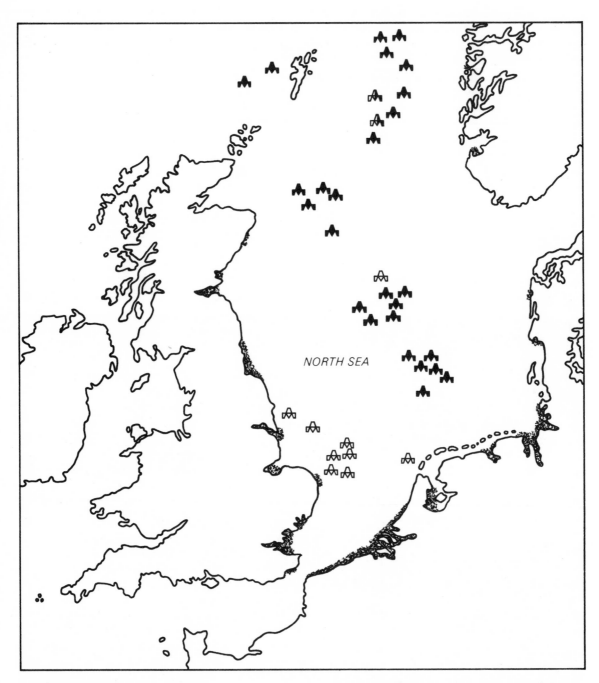

NORTH SEA

RESOURCE AND POLLUTION

The major North Sea oil and gas fields, and the chief areas of pollution from the effluents and industrial wastes of bordering countries.

- oil field
- gas field
- chief areas of pollution

once they have settled out they alter the type of substratum and hence the life of the animals living in or on the bottom. They may consist of suspended particles of organic matter, good food for suspension and detritus feeders alike. However, more often than not, inmixed with the particulate food is much inorganic material which is intractable to the food capturing, sorting, and/or digestive systems of the animals. This material will in time come out of suspension to settle on the bottom where it may clog and block up the myriad holes between the natural particles forming the sea bed. In this way the interstitial fauna, that is, the complex of minute animals living among the particles, will be radically altered if not totally eradicated. The cloggers can also wreak havoc with the food-sorting mechanisms of the suspension and detritus feeders.

Finally, and most ominous, there are the potential killers, the poisons and toxicoids. Some such as lead, copper, cadmium, mercury, and even arsenic are natural products of the earth's crust. They are always present for they are dissolved in the sea, although usually in such minute quantities that they are harmless (unless you happen to drink a phenomenal amount of water and selectively extract and absorb the nasties from it).

Oil is of course a natural product and ever since it was formed and trapped in the rocky reservoirs, erosion and geological upheavals must have resulted in leakages into the sea resulting in natural pollution hotspots. It is a suspender, a clogger, and a killer, but it is also an enricher for there are a number of marine bacteria that can feed on fats and oils as long as they are given sufficient time and the right conditions. In the same way there are many types of organism with a natural in-built tolerance of some of the toxicoids voided in high concentrations into the sea by man. For example, there are populations of rag worms living in sediments naturally rich in lead that have evolved such a tolerance to it that some of their tissues contain very high levels of what to the average rag worm is a lethal substance. Similarly

sardines found in the Tyrrhenian Sea, from whose Mediterranean shores most of the world's mercury is mined, can contain in their living tissues much higher levels of this very toxic element than are found in sardines fished from other areas. These and many other examples make it clear that the natural system can adapt to overcome at least some of man's environmental misdeeds.

However, there is a much more insidious way that man is beginning to affect the fitness of the oceans. He is today both a geomorphological and a geochemical force. Not only is he removing, transporting, and concentrating the natural products of the earth's crust but he is manufacturing new chemicals, compounds completely alien to this planet and to the process of evolution, chemicals that are unfit in the evolutionary sense. Some of them, like DDT, aldrin, dieldrin, and a whole range of others, are designed to be specifically toxic to certain forms of life; others, like the PCBs (polychlorinated biphenyls), are in reality toxic accidents, by-products of the plastics industry, which are released in vast quantities into the atmosphere and the sea and change the fitness of the environment.

Exactly what is the effect of this fricassee of change? Is man really polluting the oceans? If he is what better place to look for the signs of change than along the north-east coast of England where industrialisation had its beginnings and has continued its expansion almost unchecked up to the present day.

The inshore marine fringe with its distinct zonation of environmental conditions is one of the most diverse habitats of life on earth. Wherever there are rock outcrops, the ecosystems present are dominated by large seaweeds, green, brown, and red. Their distribution is zoned in respect to the environmental conditions. The most extensive zones are those dominated by the brown kelps, the smooth-stalked kelp just below low water mark, and the rough-stalked kelp extending to depths of up to 30 metres.

The productivity of the kelp beds is high and even at depths of 10 metres the forest may net more than 10 grams per square metre per day. The rough-stalked kelp, *Laminaria hyperborea*, is a perennial, its root-like holdfast and stiff stalk growing for periods of up to fifteen years. Just like a tree on the adjacent land masses, it grows faster in the bright warm days of spring than it does in the dull cooler winter days, and this differential helps to accentuate internal growth lines, the number of which, interpreted with care, may give the age of the kelp 'tree'. A new leaf-like lamina is produced each year and as it expands the old frond erodes away. The particulate matter so produced helps to feed the rich fauna living in the shelter of the kelp forest.

Apart from insects, birds, amphibia, and reptiles, these underwater forests provide shelter for an almost complete cross-section of the animal kingdom and a safe and firm foothold for a host of other seaweeds, some large but the majority very small. Bacteria and fungi live in the surface slime that lubricates the fronds allowing them to slide undamaged over the rocks. Protists and other microscopic plants grow on the surface of the kelp, while long chains of filamentous diatoms may cover parts of the older blade with a thick brown felt. The holdfast and stipe are often covered with brightly coloured sponges that vie for space with encrusting seaweeds, branching hydroids, stiff jelly-like colonies of sea squirts, and the regular repeating pattern of bryozoans. In the interstices of the branches of the kelp's holdfast, worms, starfish, sea urchins, and molluscs of all sorts may be found in abundance, together with sea spiders, errant stars, and a whole range of tiny fish. Pipe fish weave their way among the stipes, and ordered shoals of fish fry do their formation hydrobatics among the fronds.

Of all the animals teeming in the protection of the forest, few feed directly by eating the tissue of the adult plants. Many like the sea urchins and univalve molluscs actively browse the epiphytes from the surface of the kelp, but the plant itself appears to be sacrosanct, at least once it has attained its maturity. There are, however, some molluscs that do feed on the kelp tissue by actively boring into the base of the stipe and excavating a living shelter as they go. The end result is that the kelp 'tree' is felled and floats away to be tossed dead and drying by the waves onto the shore. The percentage of infestation of the kelps is usually low and only the oldest individuals succumb to what must be regarded as a timely demise. It would thus seem that the only animal that evolved to feed on the rich abundance of the kelp beds is, or rather was, Steller's sea-cow, now but a memory in zoology textbooks.

The majority of the herbivorous animals living in the kelp forest appear to exist in some form of truce with the productive, protective kelp. The whole forest is a structured living system closely resembling its dry land counterpart in that its integrity is maintained only as long as regeneration is possible, and that is the problem. Although the adult kelp can live with little fear of being eaten, the same is not true for the intermediate stages in its life history, for many herbivores actively graze the rocks on which the seedlings and sporelings grow.

This fact was brought home in rather an unexpected way during the aftermath of *Torrey Canyon*. In the areas of coast blanketed by the oil there was little visible or measurable effect of contamination below low water mark. In contrast, in all the areas where detergent had been used to emulsify the offending oil, a surprising thing happened. The rocks soon became covered with enormous numbers of young kelp plants. Where previously only one or two would be found waiting for a gap in the canopy to light their way to adulthood, there were now twenty or thirty—*quod erat demonstrandum*, absolute proof that baby kelp plants like being washed with detergent, because detergent kills off many of the animals that normally browse the glades of the kelp forest. Due to intense competition for light in the dense understorey, the balance of the forests was soon reinstated and two years later it was impossible to pick out the affected forests from their unpolluted neigh-

Sea-Quest, a new island in the North Sea: new potential for industrial man.

bours. Here then was proof of the existence of a dynamic system, a complex of organisms in balance with one another and with their total environment, a system that could be used to monitor the potential effects of pollution.

Along the north-east coast of England and the south-east coast of Scotland kelp forests are widespread features of the sub-littoral. The presence of these sub-aqua wonderlands are belied by the crowded fronds nodding from stiff stalks during periods of low spring tides, and the tangled masses of older individuals washed up onto the beach during stormy weather. Down this coast a pollution gradient exists, from the clean, clear waters of Scotland to the turbid, long-polluted waters of County Durham, the centre of the industrial conurbations of the north-east.

In 1968 a study was instigated by the British Sub Aqua Club, which claims the title of the largest diving club in the world. Its aim was to determine the *status quo* of marine pollution along the east-coast gradient, using the kelp forest as a system for monitoring the health of the inshore environment. The study showed that wherever the kelp was growing in polluted water its depth range was very restricted. The extensive forests typical of unpolluted sites were not found in any of the polluted areas; instead there was no more than a narrow belt of it just below low water mark. This restriction in depth range could have been brought about by a number of factors, the most obvious being that the pollutant in suspension was cutting down the amount of light penetrating the water. However, as at least some of the pollutant is a potential alternative source of food for certain animals, an increase in the abundance of browsers could not be ruled out. Was it too little light or too many grazing animals that restricted the kelp to the shallows?

Measurement of the performance of the individual plants helped to solve the problem. At each site, whether pristine or polluted, the plants growing at the greatest depth were con-

A blue-rayed limpet, the diminutive eater of kelp.

siderably smaller than those growing in shallower water, a good indication that their depth range is limited by lack of light and not by grazing. The study thus showed that pollution produces dirty water which in turn limits growth —hardly a surprising result, but that was not all.

Using the considerable potential of BSAC's army of divers, the populations of the animals that seek shelter in the holdfasts of the kelp were studied and a census was taken at regular intervals over a period of four years. A clear pattern began to emerge: the fauna of the polluted holdfasts were dominated by suspension feeders, mainly mussels and worms, which filled the house in the holdfast to the exclusion of almost all others. In contrast, unpolluted holdfasts sheltered a much more diverse population of omnivores, carnivores, herbivores, and suspension feeders. The reason for this simplification under the stress of pollution is, in all probability, that the excess of organic material, waste to man but good food to a hungry suspension feeder, puts them at an enormous advantage over all the other members of the natural ecosystem.

Given the pollution, the opportunist suspension feeders are quick to exploit the potential— the shelter is there and so the niche is filled to overflowing. Due to the abundance of man-made 'food' the simple food chain of, for example, sea stars living on a surfeit of mussels has replaced the normal more complex food web. However as more polluted sites were studied, more animals were added to the list; indeed, in time it was possible to find the majority of the animals that were present in the adjacent unpolluted systems, but in very small numbers. Evidently they can all still live in the polluted waters but are squeezed out by the super-abundance of the suspension feeders. The house in the holdfast is full, monopolised by a few types of animals to the exclusion of almost all others.

Therefore it would appear that the kelp holdfast system, placed under a 'stress for some, unbounded opportunity for others' situation, has responded by producing what is in effect a highly efficient sewage filtration plant. But everything

The animals removed from a known volume of space in the holdfasts of kelp plants: an unpolluted site has lots of different sorts of animals (above), and a polluted 'forest' has only mussels, worms, and sea stars (right).

in the effluent garden is not altogether lovely, for there are signs that the newly evolved system is unstable. The populations of the suspension feeders, and to a lesser extent that of their predators, fluctuate wildly. The holdfast habitat can be emptied virtually overnight to be refilled by new organisms as larvae become available. Such massive fluctuations are typical of dense populations of single organisms such as those found in estuarine situations where the causes, although obscure, are quite natural. The work is continuing in an attempt to explain these population crashes, to find out whether it is heavy metals, DDT, PCB, or some other doom watch toxicant

about which we have no knowledge, or simply the natural culling of overcrowded monocultures.

The most surprising result of the BSAC's investigations is that despite the long history of pollution by both industrial and domestic waste, the measurable effects along the north-east coast of England appear to be minimal. There are stretches of the North Sea, especially those in the most highly industrialised estuaries, where the muds are so polluted that they are devoid of oxygen and devoid of life. But then so too are there stretches of unpolluted fjords where the deep, still waters are also anoxic and lifeless; and there are areas below the rookeries of nesting seabirds where the kelp forests harbour a fauna of suspension feeders rejoicing not in human waste but in the surfeit of guano.

Hot spots of pollution, stretches of beach fouled by excreta, toilet paper, plastic, and all manner of man-made materials, can and should be controlled, but effective control costs an awful lot of money. So perhaps the wisest course of action would be to accept the hot spots, at least in the short term, but work to reduce them and eventually eliminate them as the economies of the world become more stable and, perhaps just as important, when we have gained a fuller understanding of the real problems of marine pollution. It would indeed be foolish to spend vast sums of money on solving a problem that is not yet fully understood.

It is essential to keep a watchful eye on the marine environment, to monitor the system so that as any signs of change become evident, immediate action can be taken to alleviate the problem before irreparable damage is done. Lessons have already been learned from accidents like *Torrey Canyon*; from situations like that at Minimata Bay in Japan, where a number of people died a lingering death and others were left maimed because they ate mussels contaminated by mercury waste; and from the Bay of Naples where an outbreak of cholera was traced to another suspension feeding organism. It is no good saying in hindsight that they should never have happened. What is important is to ensure that they do not happen again.

In the meantime the 'natural' systems are doing their best to clean up our environment, and the bottom fauna are building their homes safe inside the ubiquitous empire of coke cans on which the abyssal sun never sets.

Man's record in the development of the North Sea resource is not so very bad. His misdemeanours relate to the destruction of rich estuarine oyster-beds, over-exploitation of the populations of gourmet delicacies such as lobster and scallop, and the creation of restricted hot spots of pollution. However, there is no getting away from the fact that if exploitation of the fish stocks goes on increasing at its present rate, they must eventually crash. When they do it will be no use laying the blame at pollution's door, that is unless you include 'too many men taking too much out' within the definition of pollution.

The lesson of the cold North Sea is that man is subject to the same restrictions as those of any other member of the evolving milieu. He must use the concept of conservation to gain for himself the title of fitness and hence the reward of survival. As the rich oil and gas fields are put into production, there will doubtless be accidents, and this new part of the marine resource will create many problems. Yet it may bring with it some of the answers. In the light of past experience, perhaps some of the money will be used to maintain the fitness of the environment, fit for all the diverse members of the North Sea ecosystem, each one of which, including man, has an equal right to its fair share of the resource.

Epilogue: the road from Castellabate

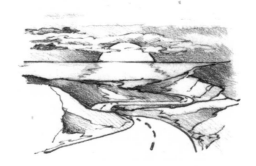

Santa Maria di Castellabate is a small fishing village south of the great Gulf of Salerno on the west coast of Italy. Its location is south of the main meccas of tourism, of Pompeii where people come to savour the sights of catastrophe, and of the Temples of Paestum that were raised in thanks for the abundant supply of fresh water that issues forth from the natural limestone reservoirs. It also lies south of the main effects of twentieth century industrial Italy. Castellabate has nothing special to offer the visitor except clean, unpolluted water and the hot inviting sun of the Mediterranean.

It was here in the summer of 1970 that a group of students, drawn together by the enthusiasm of Peter Dohrn, grandson of Anton Dohrn who in 1873 founded the marine biological station at Naples, and sponsored by UNESCO and MAMBO (Mediterranean Association of Marine Biologists and Oceanographers), worked together to study the potential of the coastline as an underwater nature reserve. Their detailed survey showed little of special biological interest: no outstanding feature, no concentration of rare plants or animals; just a piece of typical coastline whose resource had been enjoyed by man since the dawn of civilisation on the edge of this land-locked sea. Our work did, however, highlight a very real problem of man in relation to that resource, the problem of overfishing.

This survey laid the foundation of a workshop in which students from eleven countries took part, some of them biologists, some economists, others just with an interest in the sea or in life in

The spoils of overfishing: the fruits of one night's labour by one fishing boat out of Castellabate. Value—not enough to pay for the fuel.

Sardine fishing off the Senegal coast.

general. Together we were presented with the problem and together we sought a solution. The result of our energies and inspiration was engendered in the concept of a natural park, and the idea was put to a meeting of international scientists and conservationists held at Castella-bate two years later. The following paragraphs are the essence of our paper given to the Convegno Internazionale Parchi Costieri Mediter-ranei. Although the viability of the Castellabate natural park is still held in political balance, the spirit engendered by that workshop lives on, and a summary of its findings makes a fitting epilogue to this book on the life-giving sea.

One of the many 'firsts' attributable to Leonardo da Vinci is an ecological statement indicating concern over the state of the Mediterranean waters. 'Thus you make a model of the Mediterranean Sea . . . In this model let the rivers be commensurate with the size and outline of the sea. Then by experimental observation of the streams of water, you will learn what they carry away, and of things covered and not covered by water. And you will let the waters of the Nile, Don, Po, and other rivers of that size flow into the sea, which will have its outlet through the straits of Gibraltar . . . In this way you will see whence the water currents take objects and whence they deposit them.'

It has taken nearly five hundred years of social evolution for man to respond to this statement, or at least to the signs that gave rise for concern. His answer has been the concept of conservation. Although the ecological model is still only in the embryonic drawing-board stage, the concept has itself evolved from the negative, static attitude of preservation into a positive force based on the sensible utilisation of all natural resources. This is the real meaning of conservation—the only facet in the social evolution of man that may gain him survival in the evolutionary sense.

All the evidence indicates that evolution had its beginnings within the hydrosphere, and that

307

Throwing an otter trawl.

the first major environment whose potential was to be exploited by life was the sea. The potential of the open oceans is limited by two interrelated factors, light and nutrients, that together control the depth of the euphotic column, which in turn controls the productivity of the living system. The standing crop, and hence the depth to which light can penetrate, is directly dependent on the abundance of the key, rate-limiting nutrients such as phosphate and nitrate, so the process of organic evolution works within these limitations but to the goal of fuller utilisation of the potential. One factor that limits this potential is the force of gravity, in that it causes a continual loss of nutrients, locked up in dead organic matter, down to the non-productive darkness of the abyss. This is, at least in the short term, a one-way system, as areas in which deep-water upwelling refertilises the euphotic zone are limited to only 0.1 per cent of the total oceans.

The evolved sustained production of the euphotic column lies somewhere between 0.25 and 0.75 grams of carbon per square metre per day. Inshore, especially where benthic ecosystems terminate the euphotic column, the limitations are less, as a good supply of nutrients is maintained by the continual erosion from the

adjacent land masses. Here, in the sub-littoral environment (which itself is less than one per cent of the total area of the oceans), three major types of highly productive ecosystems have evolved. In the temperate latitudes the megaphytic algal systems have a net production of dry matter of around 10 grams per square metre per day. In the clearer tropical waters their place is taken by the more complex coral reef systems netting 35, and between, and interdigitating with the two main types, are the turtle grass meadows with a net production potential of 25.

Within these systems the organisms and their performance are zoned mainly in respect to the quantity and quality of the incident light energy. Each ecosystem is therefore structured in both space and time in relation to the potential of the environment, and each is just as much a product of the process of evolution as are their component species. In the main surge of evolution, certain species left the sea to utilise the potential of the dry land. Some products of this terrestrial evolution returned to the untapped resources of the marine environment to become the sea grasses and the marine mammals, for the golden rule of evolution is that *wherever there is potential the evolutionary process will respond and utilise it.*

Two sides of a resource: a young sea
fisherman and a new oil refinery.

This is the other side of the evolutionary coin: not a struggle for survival of the fittest, but the evolution of an integrated living system that mirrors the potential of the environment. No other habitat on earth demonstrates this as forcibly as the sea. Only there can an almost complete cross-section of the animal and plant kingdom be found, including a staggering variety of invertebrates and algae, groups that appeared early in evolution and, having survived this long, can be said to bear the Darwinian stamp of success.

Although the source of the evolutionary effort was the oceans, once the major step onto dry land had been taken, the two main components of the biosphere, being the hydrosphere and the lithosphere, became biologically distinct. The abiotic factors linking the two are of enormous importance and centre on the hydrological cycle, but the biotic links are limited to maritime and migrating birds feeding on fish, a limited amount of terrestrial carnivore activity in relation to those fish that return to rivers to spawn, haul outs of seals and their kin, polar bears, and the suicidal tendencies of the lemming.

The evolution of man has changed this, opening up a massive biotic link between the two so

that the world fish stocks are today endangered and the fitness of the oceans themselves may be in jeopardy. It is interesting to look at the ways in which maritime man has evolved in relation to his immediately accessible marine resource: the variety of boats, of fishing tackle, and of fishing methods used, and the knowledge of local weather and sea conditions and of the biology of these species of commerce. Wherever the indigenous fishing population attempts to supply the needs of a population larger than their own evolved society, the local biotic resource has suffered. For the fact is that the world's fishing methods are tending to become more and more standardised, leading towards the modern fleet that depends more and more on technology and less and less on the evolved skills of the men themselves.

It was in the summer of 1970 that the international workshop was set up at Castellabate to try to understand these relationships by using the local fishing ecosystem as their unit of study. The problem was apparently simple: the biotic resource was smashed, the whole section of coast overfished. The fishermen were no longer teaching their skills to their children and were in fact campaigning for them to go into other

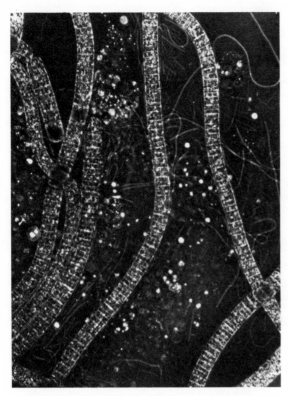

Part of a marine ecosystem: blue-green algae fixing nitrogen, and bacteria helping to break down the waste (magnification × 475).

trades and professions. The whole evolved, structured system was in imminent danger of extinction.

A solution was sought and found in the concept of the natural park, where the totality of the resource would be conserved; the key decision was made by the fishermen themselves, namely to restrict fishing in certain areas of the park, thus allowing rehabilitation of the fish stocks. The workshop's initial aim was to maintain the skills of the fishermen within the new structure of the park ecosystem by integrating their knowledge and expertise into new uses of the resource, such as recreation, education, and mariculture. Its long term aim was to re-create the complex living system with man once more part of that system, and the emphasis was on the development of a fisherman's co-operative that would control and draw benefit from the park's resource.

The workshop emphasised that an important function of the sea is its ability to dilute, disperse, and dispose of the many products of modern technological man, and it accepted that man cannot stop using the sea as a sink but stated that such use must in the future be carefully monitored. The results of the workshop were summarised in the seventeen chapters of *Plan Castellabate*. Here in this document is the essence of a blueprint for survival, not only of Castellabate and its highly skilled fishermen, but for the whole of the Mediterranean.

When the workshop members returned to their various countries of origin, each began to work to the same goal of the creation of natural parks, and already there is positive action from six Mediterranean countries. The reason that these parks are called natural and not national is simple. The marine reserve knows no bounds. The sea is the arena of evolution and each park will represent a small part of that great natural system, each a working unit of evolution. The fact that our understanding of evolution is still in the data-gathering stage emphasises the importance of the natural parks as data banks in the world matrix of genetic information. If any part of this data is lost, the process and our understanding of it will be impaired.

Anton Dohrn founded the marine biological station at Naples in order to allow Darwin's theory of evolution to be put to the test in the sea. In the same way that the Naples station has demonstrated for more than a hundred years the importance of international co-operation in research, this workshop demonstrates the same importance of international co-operation in conservation. And just as the work at Naples has been vital in our understanding of the marine resource, so too is this workshop vital to the future of the great resource that was the subject of Anton Dohrn's interest, namely the Mediterranean, one very small part of the life-giving sea.

opposite
A productive unit of evolution. Rocky headlands surround a small bay, and a local fisherman tends his boat.

Metric conversions

Except for temperature, the non-metric equivalents of the metric values 1 to 10 are given. To convert these to tens, hundreds, thousands etc, simply move the decimal point one, two, or three places to the right; eg 5 km = 3.11 miles, therefore 500 km = 311 miles. Approximations can be made: 96 km is in between 90 km (56 miles) and 100 km (62 miles), ie about 59 miles. If greater accuracy is required, add together the values of the component numbers, thus 96 (90 + 6) would be 55.9 miles + 3.73 miles = 59.63 miles.

Length

10 millimetres = 1 centimetre
10 centimetres = 1 decimetre
10 decimetres = 1 metre
1000 metres = 1 kilometre

centimetres	inches
1	0.39
2	0.79
3	1.18
4	1.57
5	1.97
6	2.36
7	2.76
8	3.15
9	3.54

metres	feet
1	3.28
2	6.56
3	9.84
4	13.12
5	16.40
6	19.69
7	22.97
8	26.25
9	29.53

kilometres	miles
1	0.62
2	1.24
3	1.86
4	2.49
5	3.11
6	3.73
7	4.35
8	4.97
9	5.59

Area

1 square centimetre = 0.155 square inches
1 square kilometre = 10.764 square feet

square kilometres	square miles
1	0.39
2	0.77
3	1.16
4	1.54
5	1.93
6	2.32
7	2.70
8	3.09
9	3.47

Mass

1000 grams = 1 kilogram
1000 kilograms = 1 tonne (metric ton)

grams	ounces
1	0.04
2	0.07
3	0.11
4	0.14
5	0.18
6	0.21
7	0.25
8	0.28
9	0.32

kilograms	pounds
1	2.2
2	4.4
3	6.6
4	8.8
5	11.0
6	13.2
7	15.4
8	17.6
9	19.8

tonnes	tons
1	0.98
2	1.97
3	2.95
4	3.94
5	4.92
6	5.91
7	6.89
8	7.87
9	8.85

Volume

1 cubic centimetre = 0.061 cubic inches
1 cubic metre = 1.308 cubic yards = 35.315 cubic feet

cubic kilometres	cubic miles
1	0.24
2	0.48
3	0.72
4	0.96
5	1.20
6	1.44
7	1.68
8	1.92
9	2.16

1,000,000,000,000 litres = 1 cubic kilometre

litres	UK pints	US pints
1	1.76	2.11
2	3.52	4.22
3	5.28	6.34
4	7.04	8.45
5	8.80	10.56
6	10.56	12.67
7	12.32	14.78
8	14.08	16.90
9	15.84	19.01

litres	UK gallons	US gallons
1	0.22	0.26
2	0.44	0.53
3	0.66	0.79
4	0.88	1.06
5	1.10	1.32
6	1.32	1.58
7	1.54	1.85
8	1.76	2.11
9	1.98	2.38

Temperature

°Centigrade	°Fahrenheit
−40	−40
−30	−22
−20	− 4
−10	14
0	32
10	50
20	68
30	86
40	104

Bibliography

The following list does not attempt to be exhaustive; it is just a number of books about the sea that have given me both enjoyment and masses of information.

Journal of Researches during the Voyage of the Beagle Charles Darwin (UK and USA, Ward Lock & Co 1891).
Diary of the Voyage of HMS Beagle Charles Darwin (UK, Cambridge University Press 1933; USA, Nora Barlow, ed., Kraus Reprint 1969).
Geology of the Voyage of the Beagle: 1 Corals, 2 Volcanic Islands, 3 South American Geology Charles Darwin (UK, Smith Elder & Co 1842–6).
Structure and Distribution of Coral Reefs Charles Darwin (UK and USA, University of California Press 1962).
Voyage of the Beagle Charles Darwin (UK and USA, Bantam Books 1973 and 1972).
Origin of Species Charles Darwin; J. W. Burrow, ed. (UK and USA, Penguin 1970).

The first five books allow you to follow the discoveries made by Charles Darwin on the epic voyage of HMS *Beagle*. They are the writings of a true natural historian, someone who attempted to integrate the total information contained within the living landscape. Out of his understanding came the concept of evolution. Although the early editions of these books are now out of print, a few are still available on the second-hand book market, and when reading a classic work, it is more exciting to read it in its original form.

Story of the Earth (UK, HMSO 1973, for the Institute of Geological Sciences, available from the Geological Museum, London).

The idea of a museum of rock may well conjure up a dreary and dusty picture, but not so the Geological Museum. It depicts the story of our planet, a story brimming over with the fire of molten rock, the shock of earthquakes, the pressures of sedimentation, and the promise of the evolution of life—all summed up in this publication.

Introduction to Marine Botany E. Y. Dawson (UK and USA, Holt, Rinehart & Winston 1966).

This covers the whole field of the plant kingdom under the sea.

Biology of Marine Algae A. D. Boney (UK, Hutchinson 1966; USA, Hillary House 1966).

All you want to know about seaweeds, and a lot more.

Animals without Backbones (2 vols) Ralph Buchsbaum (UK, Penguin 1971; USA, University of Chicago Press 1972).

Here are two books which, apart from being definitive works in their own right, read like novels. A word of warning: don't start them if you haven't time to finish the job, for they are compulsive reading.

Marvellous Animals Helena Curtis (UK, Heinemann 1969; USA, Natural History Press 1968).

The protists were first described in detail by Antony van Leeuwenhoek in the mid seventeenth century, and he christened them 'animalicules' or little animals.

Annelids R. P. Dales (UK and USA, Hutchinson 1967 and 1963)

The titles in the Hutchinson University Library series are made to fit into your pocket and are written to be read again and again. This one is well worth worming your way through.

Molluscs John E. Morton (UK and USA, Hutchinson 1967).

A super little book about a super group of animals. Once you have read it and found out all about their ways of life, I am sure you will agree that one mollusc alive in the sea is worth ten thousand shells on the mantlepiece.

Oysters C. M. Yonge (UK and USA, Collins 1960).

An intimate account of the private life of a very special mollusc.

Life of Fishes Norman B. Marshall (UK, Weidenfeld & Nicholson 1965; USA, Universe Books 1966).

A book full of magic. It not only tells you everything you would like to know about the fish, but answers many questions that I am sure the fish themselves would like to ask.

Old Fourlegs J. L. B. Smith (UK, Pan Books 1956). The complete unabridged story of the discovery of the coelacanth, one of our links with the past.

Birds of the Ocean Wilfred B. Alexander (UK and USA, Putnam 1955 and 1963).

The sub-title of this book is 'A Handbook for Voyagers', and although somewhat out of date, it is well worth taking on any voyage.

Sea-birds James Fisher and R. M. Lockley (UK, Collins 1954).

This makes great reading, is brim full of important detail, and has many wonderful pictures.

Biology of Marine Mammals H. T. Anderson, ed. (UK and USA, Academic Press 1969 and 1968).

The definitive work by thirteen world experts. Although tough going in places, it contains the facts about underwater mammals.

Half Mile Down (revised edn) Charles William Beebe (UK, Bodley Head 1951; USA Hawthorn Books 1951).

Man will conquer inner space—a classic by a classic pioneer of underwater exploration.

The Silent World Jacques-Yves Cousteau and Frederic Dumas (UK, Hamish Hamilton 1953; USA, Harper & Row 1953).

A personal account of our other world, written by the man who put man among the fish.

Open Sea: Its Natural History (2 vols) Sir Alister Hardy (UK, Collins 1971; USA, Houghton Mifflin 1971).

Volume 1 weaves the intricate story of the world of plankton, while volume 2 details the interdependence of the fish and hence our fisheries on the productivity of the plankton soup.

Great Waters Sir Alister Hardy (UK, Collins 1967; USA, Harper & Row 1967).

'A Voyage of natural history to study whales, plankton, and the waters of the Southern Ocean in the old Royal Research Ship *Discovery*.' This account puts the sound of the living sea right on your bookshelf.

The Sea Around Us Rachel Carson (UK, Panther 1969; USA, Oxford University Press 1961).
The Edge of the Sea Rachel Carson (UK, Panther 1973; USA, New American Library 1971).

By the author who made the world aware of the growing problems of pollution, these two books live like the sea they describe.

Oceanography: Readings from Scientific American J. R. Moore, compiler (USA, W. H. Freeman 1971).

Every month *Scientific American* brings the best of science to your news-stands. Here is the best from the best about oceanography, bound together in what can only be called a supervolume.

Illustration credits

Page numbers given, those in italics refer to colour.

Heather Angel: *18*, *35*, *36*, 46, 48, 49, 50, 54 (bottom left), 55, 56, 57, 69, 70, 71, 77, 78, 79, 80, 81, 82, 83, 86, 87, 99, 100, 107 (both), 111 (both), 113, 116, 118, 120, 121, 123, 124, 125, 126, 129 (bottom), 132 (bottom), 133, 134, 136, 139 (both), 141 (both), 142, 143, 144, 148, 149, 150, 151, 153, 154, 157, 158, 159, 160, 161, 168 (both), 169, 174 (both), 180, 183, 186, 188, 190, 191, 193, 195, 197, 199 (bottom), 200 (both), 203, 214, 215, 223, 246, 247, 250, 259, 267 (both), 270 (both), *272*, 273, 275, 276, 277, 302, 307, 308, 309.

Ardea Photographics: H. & J. Beste 238; R. J. C. Blewitt 285; J. B. & S. Bottomley 269, 287 (both); Ian Curphy 245; K. Fink 240 (bottom); S. Gooders 32 (both); Clem Haagner 230, 233, *235*; E. Mickleburgh *17*, *236*, 242, *254*, 256; P. Morris 19, *73*; Bryan L. Sage 268; P. Steyn 228; B. Stonehouse 229; R. & V. Taylor *62*, 201, *207*, 210 (bottom), 226, 240 (top); R. Vaughan 286; A. Warren 237, 239.

Australian Information Service, London: 29.

David Bellamy: 51, *64*, 234, *265*, *271*, 284, 304 (both), 305, 306, 311.

J. Barnes: *218* (both).

British Petroleum Company Limited, London: 301.

British Tourist Authority: 290, 292.

Bruce Coleman Limited: Jane Burton 59, 137; B. Campbell 227; Bruce Coleman 221; J. Dermid 88 (right); J. Foott 241; D. Hughes 130; Oxford Scientific Films; *91*, *109*, 184; G. D. Plage 212; A. Power 262.

B. M. Davidson: 41 (top), 44, 76, 88 (left), 115, 122, 129 (top), 138, 185, endpaper.

Peter Dohrn: 8, 9.

R. Harding Associates: 20; R. Cundy 255 (both).

M. J. D. Hirons: 231, 232, 294.

D. J. Hutchinson, Photographic Unit, Department of Zoology, Durham University: 68, 162, 171.

Imperial Foods Limited, Grimsby, Lincolnshire: 274.

Japan National Tourist Organisation, London: 278, 279.

Gordon F. Leedale: 38, 39, 310.

Museum of History of Science, Oxford University: 47.

Scarborough Zoo and Marineland, Yorkshire: 222.

George Russell: 85 (both).

Seaphot: T. Baverstock 94; D. Clarke 108; P. David *74*, *92*, *128*, 131, 152 (top), 251, *253*; W. Deas *61*, *127*, *163*, *208*; C. Doeg 27; Werner Frei 117, 196, 219; J. David George 65, 66 (all), 67, 105 (both), 106, 165 (bottom), 166 (both), 175, 199 (top); G. Harwood 54 (top right), 84, 152 (bottom), 165 (top), 198, 202, 259, 295; Dr D. Hibbard 45; Gordon F. Leedale 41, 42, 43, 72; John Lythgoe 89; D. Guiterman 296; C. Petron *110*, *145*, *146*, 147 (bottom), 170, 176, 194, *205*, *206*, 210 (top left), *217*, 225 (bottom); R. Salm 132 (top); P. Scoones *63*, 147 (top), *164*, 167, 209, 225 (top); P. Smith 14, 15, 192; P. Vine, Coral Conservation Trust 266.

A. Schmidecker: 181.

United Kingdom Atomic Energy Authority: 297.

The line diagrams are by Hilary Evans.

The drawings are by Amaryllis May.

Index

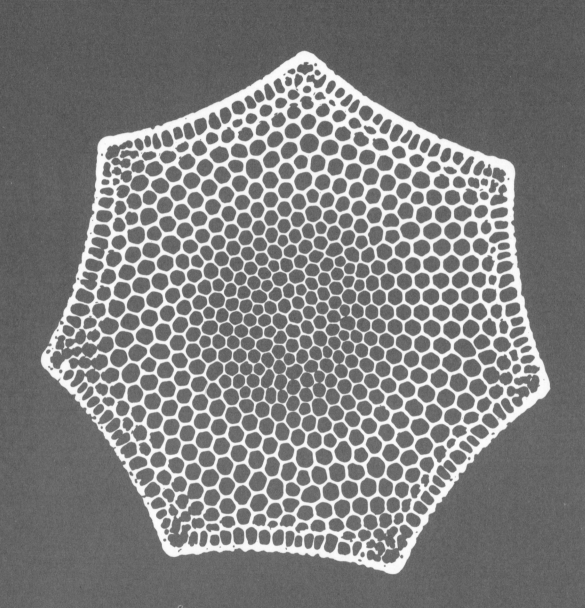